Edward Sylvester Ellis

A Jaunt Through Java

The Story of a Journey to the Sacred Mountain by two American Boys

Edward Sylvester Ellis

A Jaunt Through Java
The Story of a Journey to the Sacred Mountain by two American Boys

ISBN/EAN: 9783744798105

Printed in Europe, USA, Canada, Australia, Japan

Cover: Foto ©Andreas Hilbeck / pixelio.de

More available books at **www.hansebooks.com**

A JAUNT THROUGH JAVA.

The Story of a Journey to the Sacred Mountain by Two American Boys.

By EDWARD S. ELLIS,

Author of

"Adrift in the Wilds," "A Young Hero," "Log Cabin Series," etc., etc.

ILLUSTRATED.

A. L. BURT COMPANY,
PUBLISHERS,   NEW YORK.

A JAUNT THROUGH JAVA.

CHAPTER I.

TWO YOUNG TRAVELERS.

"HURRAH for Java!" shouted Hermon Hadley, flinging his cap in air, and uttering a shout that might have been heard a half mile, except for the parlor walls which shut in the sound.

"I'm with you!" added his cousin Eustace, hurling his hat against the ceiling and executing a sort of double-shuffle on the carpet, while the mothers of the lads tried to look stern and severe, but were unable to keep back their smiles.

"We should be thankful for their rugged health and strength and high spirits," remarked the mother of Hermon, "for if those ringing voices were hushed in death how we would miss them, sister!"

Tears filled the eyes of both, which quickly gave way to laughter at the antics of the youngsters, who disported themselves like a pair of monkeys. Eustace dropped upon the stool in front of the piano and rattled off a jig, while Hermon executed it with commendable vigor, even if he did not keep the best of time. After a while the boys were able to sit down and calmly discuss the important business with their mothers, who were sisters, while their fathers were brothers.

Hulbert Hadley lived in the tropical island of Java, while his brother Isaac was a merchant in St. Louis, the latter being the father of Hermon and the former the parent of Eustace. When the Javanese Hadley, as he may be called, brought his wife and son from the other side of the world, they made a visit of several months to their relatives in America. During this visit the arrangement was made that Hermon should join his cousin in attending school in England.

This was well enough; but when Eustace and his parents urged that before carrying out this excellent plan the American youth should make them a visit in Java, the St. Louis merchant was not prepared to give assent. It seemed to him that the time would be wasted, and now, while his son was growing so rapidly, he ought to devote his energies to acquiring an education.

But the boys, as might have been expected, hung on, and by and by won their mothers to their side. The father of Eustace had favored the plan from the first. It came about, therefore, that when Mr. Isaac Hadley found that both households were arrayed against him he exclaimed:

"Well, I suppose it's no use! Yes; you can go to Java!"

It was this announcement, made at Mr. Hadley's store in the presence of his brother, that sent the youngsters flying home like a couple of Comanche Indians.

It isn't worth while to dwell on the particulars of the preparations made by the boys. The father of Hermon was obliged to stay in St. Louis to look after business matters, while his wife and son accompanied the other family on the long journey to the distant country about which I hope to tell you something that will prove interesting and instructive.

Eustace, having made the voyage to America by way of London, was able to give his cousin much information respecting the wonderful sights and scenes which met them on every hand. Hermon had made several visits to the metropolis of his country, where he spent nearly a week visiting the places and objects of interest. Finally, on a mild, sunshiny Saturday, they bade good-by to America and embarked on one of the magnificent Cunarders for their distant destination.

Three state-rooms had been engaged by the party—one for the parents of Eustace, another for the mother of Hermon, while the boys roomed together. The steamer moved swiftly down New York harbor, and off Sandy Hook a brief stop was made to allow the pilot to leave. Finally the Navesink Highlands sank out of sight below the horizon, affording the last glimpse of Hermon's native land.

Our friends were disappointed in their expectations of a pleasant voyage across the Atlantic, for the weather became blustery, and such a high sea was running that every one of the company except Mr. Hadley became very seasick. The lads, however, quickly recovered, and although they were quite pale and felt weak in the legs, they soon made their appearance on deck, where they were chaffed by Mr. Headley for being such poor sailors.

There were many games, several concerts, singing, story-telling, and the various amusements with which several hundred people pass away the hours on shipboard. The boys found much entertainment in standing on deck and watching the vessels, some of which were nearly always in sight, and speculating on their destination, their past and their future history and fate. Once considerable excitement was caused by the sight of a mountainous iceberg floating slowly southward. There was talk of a collision with the enormous mass of ice; but the expression of confidence on the faces of the officers and crew dissipated what fear any one might have felt, and the distance between the steamer and ice-mountain was never less than three or four miles.

On the ninth day from New York our friends landed in Liverpool. There was so much ahead of them, and they were so anxious to reach their far-away destination, that they left the great city, with its famous docks, the next day, taking the London and Northwestern Railway to London.

Here they were in the most populous city on the globe, and fully a week—altogether insufficient—was devoted to sight-seeing. The yellow, sticky fog, for which that mighty city is so famous, was over everything, and rendered it impossible to see more than a few feet in any direction; but when the weather had cleared somewhat they visited Cheapside, Fleet Street, Ludgate Hill, Pall Mall and Piccadilly, which are among the principal thoroughfares. Then followed an inspection of St. Paul's Cathedral, the Tower of London, Westminster Abbey, the Houses of Parliament, the Bank of England, known as the Old Lady of Threadneedle Street, and other interesting places.

"We could spend a month here," said Hermon, "and use every hour, day and night, in seeing the sights."

"Yes," replied Mr. Hadley, "and leave many interesting places unvisited. In fact, every country is like an immense volume of treasures which you can examine and study, and then begin again and find you have missed some of the most wonderful of them all."

Had we the space, the full account of the journey of our friends to Java, which was still thousands of miles off, would be entertaining and instructive. From London they went by way of Dover and Calais to Paris. The little packet-ship was so tossed about on the channel that all, including Mr. Hadley, had two hours of dreadful seasickness. The landing was at Calais, whence they traveled two hundred miles by rail to the beautiful and in some respects most famous city in the world.

It would never have done to pass through Paris without a hurried inspection of some of its most striking features. The gay boulevards, where, especially at evening, the whole city seemed to be on parade; the Gardens of the Tuileries; the Place de la Concorde, the most magnificent public square in the city; the Egyptian obelisk of Luxor, where over a thousand persons, including many royal personages, were beheaded nearly a hundred years ago; the Arc de Triomphe, the largest triumphal arch in the world, with the streets radiating from it like the spokes from the hub of a wheel; then, crossing the Seine by the Pont Notre Dame, the oldest bridge in Paris, to the island of La Cite, where a Gallic tribe built a hamlet before the birth of Christ, and where stands the noble Gothic Cathedral of Notre Dame, more than seven hundred years old, with its huge equestrian statue of Charlemagne; to the south side of the Seine in the Latin Quarter, where most of the institutions connected with the University of France are situated; among the Catacombs, with their ghastly remains of the myriads who have died and been forgotten; in the Jardin des Plantes, with its interesting animals; the Hotel des Invalides, with its golden dome and the tomb of Emperor Napoleon, and the almost numberless places famous in history; these were visited by the ardent travelers, despite their eagerness to push on to Java.

The time came, however, all too soon, as it seemed, for their departure from France. They left Paris for Cologne by rail, arriving in the German city on the Rhine twelve hours later. The distance is about three hundred miles, and took the travelers across Northern France, Belgium, and a portion of the German Empire.

The great attraction of Cologne is its Cathedral, whose foundation-stones were laid more than six hundred years ago, while the cap-stone was not put in place until 1880. Its towers are upward of five hundred feet high, affording a view of hundreds of miles, and making them the loftiest spires in the world excepting our own Washington Monument and the great tower now in course of erection in Paris, which is destined to be a thousand feet in height.

They left Cologne by steamer, down the picturesque Rhine, which recalls so vividly our own Hudson. At Mayence the boat was abandoned. A day's travel by rail brought them to Zurich, and thence to Geneva. Several days were spent in this interesting city near the Alps, whence the little party journeyed through the Mont Cenis Tunnel, one of the most amazing feats of engineering the world has ever known.

Two parties began work nearly eight miles apart, and thirteen years later (1870) met in the middle of the mountain. Had the engineers made only a slight miscalculation, these parties would have passed each other and completed two tunnels—that is, if the money and patience had held out.

It took ten hours to cross the kingdom of Italy from Turin to the Adriatic. A few days were delightfully spent in Venice, the next stop being at Vienna. Thence they passed by steamer down the Danube to Buda-Pesth, and another day's voyage through the "Iron Gates" brought them in sight of the fortifications of Belgrade.

We must not dwell too long, however, on the preliminary journey, as it may be called, of our young friends, though continually tempted to refer to the strange and interesting scenes they witnessed. Through Turkey by rail, three days on the Black Sea, thence by rail and steamer again, until at last they disembarked at Cairo, in Egypt.

Here at last they found themselves in the land of the Pharaohs, a country so rich with historical associations that no one of the company was willing to push on without tarrying long enough to look upon some of the wonderful curiosities, which it may be said were on every hand.

Not the least interesting was the mummy that had been disinterred but a short time before. It was that of the great Egyptian conqueror, Sesostris, who ruled Egypt more than a thousand years before the birth of our Saviour. It was preserved with startling perfection. The crown was bald, but there was a fringe of gray hair along the back and sides of the head, stained yellow from the spices of the embalmers, while the eyeless sockets seemed to stare in the faces of those who had come, thirty centuries after, to gaze upon the remains of one of the mightiest monarchs that ever reigned.

The first noteworthy visit was to the Pyramids and the Sphinx, a short distance from the city, on the Nile. The excavations were then under way about the Sphinx, and had progressed far enough to disclose the remains of a splendid temple, undoubtedly the oldest in the world. For unknown centuries the Sphinx has been buried to the neck in the sands of the desert. It is one hundred and forty feet long, and is hewn from the hardest rock. As yet no one has been able to solve the mystery of its existence.

Through the Suez Canal, along the Red Sea, across the Arabian Sea to Bombay, thence by rail to Calcutta, where our friends took passage direct down the Bay of Bengal to Java. When they entered the vast Strait of Malacca, between that peninsula and the Island of Sumatra, they crossed the boundary line between Asia and Polynesia, and thenceforward lived for a time on that continent of islands.

It was in the month of May that the little party reached Samarang, where their long wanderings were ended for the time. This was a favorable date for their visit, since it was the beginning of the dry season, which opens in April and lasts till October. During that period the prevailing wind is the southeast trade, and for the rest of the year it is the northwest or west, which is the continuation of the regular north-east trade-wind. The rainy season, though less defined in the eastern part of the island, is so uncomfortable that the visitors were fortunate in timing their arrival so that the unacclimated ones avoided it.

Samarang, when first seen by Hermon Hadley and his mother, contained about twenty-five thousand inhabitants. It had a quaint and interesting appearance to the Americans, whose stay in Batavia was too brief to allow them to go far into the interior of the island.

The old European portion of the town is an almost exact copy of a Dutch city, the sturdy Hollanders who built it seeming to fancy the climate to be precisely the same as that which prevails on the banks of the Zuyder Zee.

The wall surrounding Samarang was taken down more than half a century ago, and a fort and coast battery stand ready to protect the city against any freebooters that may come stealing over the China Sea or from among the labyrinthine Spice Islands.

"I'm glad we are here at last!" exclaimed Hermon with a sigh of relief, "and thankful that not one of us has had a day of sickness."

"How about the third day out from New York?" asked his uncle with a smile; "you seemed to be somewhat under the weather then."

"You learned how it was yourself in crossing from London to Calais," laughed the youth; "but seasickness don't count, you know."

"When you were moaning in your state-room you were sure you were never so ill in all your life, and said you thought it would be a comfort to die. However, you are right; we have a great deal to be grateful for, for it is seldom that a party of five can travel so many thousand miles without meeting with accident, as well as suffering dangerous illness."

"And we have been in some of the unhealthiest regions in the world!" added Eustace.

"Where are they?"

"I believe that it is agreed that the delta of the Nile is the breeding-ground of cholera and other abominations that have swept off thousands of people. Undoubtedly it was there that the pestilence had its birth."

"India is about as bad," remarked the mother of Eustace, "for it is the home of plague and famine.

If it was our fate to spend the hot season there, I am sure we would not survive."

"You know we have some pretty warm weather in our country," said Hermon. "Father once made a journey through Southern Arizona in a stage-coach, and he said the thermometer stood at one hundred and twelve degrees, day after day, in the shade."

"A man can't stand much more than *that* and live," was the comment of Mr. Hadley; "Isaac told me about it," and if I hadn't known him to be a truthful man I would have discredited the statement."

"It would seem, uncle, that since Java is a tropical island, you are liable to the same kind of weather here."

"Of course it is hot at times, but the ocean is on every side, and the wind is always blowing in one direction or another. Besides, although Java is a pretty fair-sized island, it is comparatively narrow, so that it has many advantages which more temperate regions lack."

"Then, I suppose, the Dutch have owned the island so long, and so many of them have lived here, that they have imparted some of the characteristics of their own country to it?"

"Unquestionably they have; but it won't do to claim that *that* fact can have produced any effect on the *climate!*"

"I did not mean to say that; but it seems as though a person can become accustomed to almost anything. You don't think there is any particular danger in Eustace and me taking a jaunt through Java, do you?"

The mothers of the boys looked anxiously at Mr. Hadley for his reply.

"Eustace has a practical knowledge of the country; and if you go across the middle of the island, instead of taking it lengthwise, and are properly equipped, with an experienced guide or attendant, there is no more danger, as I view it, than in taking a ride in a railway train."

"Then we have your consent, father?" asked Eustace.

"Oh, I suppose so!" he laughed; "for I shall have no peace until it is granted, and your mothers won't have any until you are safe back again!"

CHAPTER II.

HO, FOR THE SA. RED MOUNTAIN!

MR. HADLEY insisted that before the boys ventured on their jaunt across the Island of Java they should spend several days at home, in order, as he expressed it, that Hermon should become acclimated to the new country. The gentleman, having been absent so long from home, found many things demanding his attention in which the youths could give him great assistance. His correspondence had accumulated to such an extent that he would have been appalled but for the nimble fingers, bright intelligence and readiness of the lads in giving him all the help in their power.

"Pitch in, my lads!" said he with a smile, "and when we get things in shape you shall have your vacation!"

"We don't want it, uncle," returned Hermon, "until we have done everything we possibly can for you. You have been very kind and indulgent to us, and we shall always remember it."

"Those are my sentiments," added Eustace, who was very devoted to both of his parents. "Hermon and I had intended to volunteer our help before you said anything about it."

The gentleman was pleased with the spirit of the boys, which, however, was only what he expected from them. The mothers were able also to lend their assistance, for they were bright-witted and well-educated, so that Mr. Hadley jestingly offered both situations as his clerks.

The servants that had been left in charge of the home during the long absence of the heads of the household had been faithful; and, indeed, matters were found in a much better shape all around than was anticipated by any member of the family.

Thus it came about that just one week after the arrival of the little party in Samarang the two cousins were tramping the rugged mountain chain which extends through the middle of Java from one end to the other. The sturdy, handsome youths were clad in light but serviceable clothing, quite similar to that which a pedestrian on this side of the world would don when starting on a lengthy excursion through the country. Each carried a revolver and a rifle of the most approved pattern, with a good supply of cartridges, a light but strong blanket, and they were accompanied by Tweak, a mongrel canine belonging to Eustace, and of which he was very fond.

The youths in their broad-brimmed hats and jaunty suits made a pleasing and striking picture as they waved their friends good-by, promising them that, if nothing interfered, they would be back in the course of a week or two.

Like all sensible persons who set out on a venturesome journey they had a definite object before them. They were on their way, as has been intimated, to the Sacred Mountain, near the southern coast of the island, in the province of Bagelen. In the neighborhood of this mountain are some of the most wonderful ruins in the world. Eustace, who had traveled over a great deal of Java, had long contemplated a visit to them, but had deferred it in the hope that his cousin might be his companion on the trip.

"When I first studied geography," said Hermon, "and saw the pictures of islands, they always seemed to me like little spots of land that you could see across. I remember that I had the same idea of the oases in the Great Desert."

"My impression was similar," said Eustace; "and it was a long time before I realized that many of the islands are vast tracts of country, with widely-separated cities, containing long rivers, extensive mountains, and thousands of square miles of territory."

"Now, as to Java," continued Hermon, who naturally had studied a good deal about it, "if I had been told that it is four times as large as Holland, the country which owns it, it would have been hard to believe it; and yet it is true."

" You consider Massachusetts a pretty fair State as regards size, but it would take half a dozen of them to equal the area of Java. Did you ever notice its resemblance, Hermon, to Cuba in size and form ?"

"Yes; the similarity is striking, but Cuba isn't one-half as interesting as Java, of which I have read considerable, when I knew you lived here and I expected to make you a visit."

"It seems to me," said Eustace, thoughtfully, "that the difference between the islands is the difference between the people. The Spanish are hot-blooded, passionate, revengeful, and ever since I can remember there has been guerrilla fighting going on in Cuba; but, so far as I know, it is a long time since there was any trouble in Java, though we have eight times the population of Cuba."

"There is no country in the world where a person is safer than here. Of course when we get up among the mountains and wild regions we may come upon evil people, just as you find them in the United States, without going out in the woods to look for them. You know what a peaceful, industrious people the Dutch are; well, Java, with all its half-savage natives, is becoming Dutch clear through. The Europeans and Oriental immigrants at present make up only a seventh of the population, the rest being what we call the Javanese proper, the Sundanese and the Madurese. I have told you the distinctive features of those people and won't attempt to get off a lecture now, but must give full credit to the Dutch for the way in which they rule this country. The emperors of Java, as their ancient history calls them, used to have their capital at Kartasura, now a deserted place, only a few miles eastward of the spot where we are walking."

"When did the Dutch gain possession of the island?"

"Nearly three hundred years ago the Dutch East India Company started their trading-posts along the coast, and some time after founded Batavia. There was a good deal of fighting, but the Portuguese and English gave way to them, and in 1749 the Javan ruler turned over the sovereignty of the island to the Dutch. The British took the island in 1811, but, five years after, it was given back to the Dutch."

"I don't clearly understand the method by which the country is governed, though I have studied it considerably."

"It is quite simple: Java is divided into some twenty-three counties, as you would call them, though we know them by the name of districts. Each district is governed by a Regent or native prince, and under him are more petty rulers down to the village chiefs. With each Regent is a Dutch Resident or Assistant Resident, who is looked upon as the elder brother of the Regent. He makes recommendations to the Regent, but they are in reality orders, for they are obeyed to the letter. Then each Assistant Resident has a Controller, who is a sort of inspector of all the native rulers, and who at regular times visits every village in his district, examines the proceedings of the native courts, hears complaints, shapes up things generally, and superintends the Government plantations."

"What are they?"

"The name tells you: the Government took steps to persuade the people, through their chiefs, to give a part of their time to the cultivation of coffee, sugar, and other products. A certain fixed rate, quite low, of course, was given to the laborers clearing the ground and making the plantations. The products are sold to the Government at a small price. From the profits a percentage goes to the chiefs and the rest is divided among the workmen."

"How does the plan work?"

"Admirably; the surplus is sometimes considerable, and the natives, instead of leading lazy, vagabond lives and becoming dangerous to the rest, are industrious, orderly, well fed and decently clothed. Of course it is a good thing for the Government; but good as it is, it is much better for the people themselves."

"The soil must be excellent?"

' Nothing could be finer, as the productions themselves show; Java is the most fertile, the most productive and the most populous tropical island in the world. The vegetation grows almost to the very tops of the mountain peaks."

"You are not lacking in your volcano supply?"

"No; we can spare a few of them, if you would like to make an exchange."

"How many volcanoes have you?"

"The number is given at thirty-eight, and some of them are more than two miles high."

"I think I would feel uneasy with such neighbors."

"You would soon get used to it; we never think about them; that faint cloud away off yonder in the south-west is caused by a volcano. A few of them are in constant activity, but regular lava streams are unknown in Java. However, I have talked until you are tired of hearing me, and after going a little further we'll stop for a rest."

It should be stated that the width of Java at the middle part, where our friends were journeying, is about forty miles. At that time a railroad connected their home (Samarang) with the towns of Surakarta and Jokjokarta, the latter of which lies near the southern coast; but the boys disdained to use the railroad—it would have been too much like riding in a carriage and calling it a walk.

They moved considerably to the westward of the places named and entered the mountainous district of Kadu, which adjoins the one containing the Sacred Mountain and the ruins they had set out to examine. They were some twelve or fifteen miles north of the volcano of Sumbeng, which sent off the thin vapor that was discerned in the sky long before the volcano itself was visible.

The road was smooth and rough by turns, but they found no difficulty in making their way over it. It showed that it was old, and had been traveled by carts and carriages for many years. Winding and turning, it steadily ascended toward the mountainous regions beyond; and the temperature which Hermon had found quite oppressive earlier in the day became cool, salubrious and invigorating. He was sure the excursion would prove one of the most pleasant possible; but, as is often the case, he little dreamed of the extraordinary experiences which even then were close at hand.

CHAPTER III.

THE HUT BY THE WAYSIDE.

TWEAK, the dog which the boys had brought with them, was of mixed breed. He seemed to combine some virtues and few of the vices of his species. He was rather large, mainly black and white, and appeared to look upon the excursion with as much frolicsome anticipation as did the lads themselves.

While the cousins sauntered along, chatting together and frequently stopping to admire the beautiful landscape and scenery opening out before them, Tweak was running ahead, dashing into the wood and out again, or chasing some bird on the wing, as though he expected to beat it in a fair race. Sometimes he would be gone for half an hour at a time, and then, when his master would whistle for him, he would gallop into sight far up the road and look back upon them, as though impatient because they were so slow.

Harmon Hadley often broke out with expressions of admiration, for, although he had seen many foreign countries, with much beautiful and strange scenery, on his voyage from England, nothing charmed him so much as the wonderfully exuberant vegetation of Java by which he was inclosed. Few of the plants are deciduous, so that the whole island is forever clothed in living green, and some of the villages and smaller towns are so hidden from view by the abundant verdure that they can hardly be seen even for a short distance.

When the youths started on their journey and for a considerable while afterward, until they began steadily ascending, they were in the zone of rice fields and sugar plantations, of cocoa-nuts, cinnamon and cotton. Many parts of the coast are fringed with mangrove and palm trees, and Hermon noticed numerous small ponds, covered with lotus flowers. Earlier in the day he had caught glimpses of prairies, covered with the silvery alang grass, broken at intervals by bamboo and patches of the taller eri grass.

These are the principal characteristics of the tropical zone of Java, which extends generally to a height of two thousand feet.

Still climbing upward, the youths found the climate deliciously cool as they entered what is called the second zone, and whose upper limit is some four thousand five hundred feet. Within this region lie most of the great coffee and tea plantations, and the sugar palm and maize also abound.

The road, which they followed at a leisurely gait, wound through woods for most of the distance, among which grew the Javanese teak, the sago, and numberless varieties of palms. Now and then they struck open stretches of country, sometimes many acres in extent, and covered with grass and rocks, and occasionally crossed by a small purling stream of water. In pushing onward and upward they met now and then persons, a few on horseback, but mostly on foot. Harmon had already learned to distinguish the pure Javanese natives from the Sundanese by their darker skin, shorter stature and stouter figure.

When the young American saw in Samarang some natives from the eastern end of the island who were not only as stout but as tall as the Javanese, he was told they were Madurese. He noticed also that most of the Sundanese had their eyes set obliquely in their heads, while those of the others were straight.

All these people belong to the Malay race, with brown or black eyes, long, coarse black hair, and very few have scarcely more than the semblance of a beard. The color of the skin presents various shades of yellowish brown with a touch of olive green. A golden yellow complexion, in the eyes of the Javanese, is the perfection of female beauty.

Eustace and Hermon were advancing at a leisurely gait, when, in turning a bend in the road, they caught sight of a bamboo hut, standing in the edge of the wood, and about a rod from the highway.

"Halloo!" exclaimed the American; "there is a native house!"

"Yes; that's the hotel where we are going to stay to-night."

"Have you ever been here before?"

"Yes; father and I were up here last July to see Jurak."

"Jurak?" repeated the wondering Hermon; "whom do you mean?"

"Jurak and his wife, Myeta, live in that hut; they have a son who is married and lives near Surabaya. Jurak is one of the best workers on a coffee plantation, and he has been most valuable to father ever since I can remember. Last summer he was bitten by a snake, and everybody thought he would die. This is a good ways from our plantation, as you can see, and he spent only his Sundays at home. We did not hear of his misfortune until the second day, when we mounted our horses and hurried out here to see him. It was a long and hard ride."

"How did you find him?"

"He was a dreadful sight: he had been bitten on the ankle, and his leg was swollen until it seemed to me the skin must burst. It was discolored so that father whispered to me he believed gangrene had set in, and there was no hope for him; but one of the native physicians came in while we were there, and took him in charge. I don't know how he handled him, but he began to get better right away, and within a week he was as well as ever."

"Which goes to show that our doctors have much to learn. I suppose Jurak was pleased with your call?"

"You never saw a more grateful fellow; he really felt as thankful to us, especially to father, as he did to the doctor who managed to pull him through. Ill as he was he knew us when we walked into his hut, and his dark eyes looked the gratitude which he tried to express in words. We brought him some wine and delicacies that mother sent him. He will be glad to see us."

The bamboo hut which they approached was strongly built, as though the builder felt there were some dangers beside the elements against which he needed protection. It was only a single story in height, with two rooms, or apartments. The floor was the ground, but the sharpened bamboo which made the walls were driven so deeply into the earth that nothing could crawl through from the outside. Fire was only needed for cooking purposes, and during the dry season that was kindled on the outside beneath the shade of some trees, where a few simple utensils were at command.

Contrary to the usual custom, there was a single window on each side of the front room, and a few flowers growing around the open space surrounding it, giving a cheerful and inviting appearance to the coffee worker's home.

Just as the boys turned to leave the highway Eustace stopped short and looked around.

"I wonder where that dog is," he said; "like enough he is a mile or so up the road."

Puckering his lips, he emitted a shrill whistle, which was hardly uttered when Tweak bounded into sight only a few rods away. The intelligent animal stood still and looked knowingly at them, as if to say:

"It's all rignt; don't fret yourself about me; I'll be within call whenever I'm wanted."

The door of the hut or dwelling was open, and before the two youths reached it Myeta, the wife of Jurak, was observed standing in it and looking inquiringly at her callers. She was short and stout, with a long, dark-colored robe, gathered loosely around her figure, and bound by a sash close under the arm, so that her appearance suggested the well-known Mother Hubbard dresses of the present day. The resemblance was strengthened by the arrangement of her head-gear. The long, coarse, black hair was gathered in a large coil on the crown, around which was tied a turban, so dark that, at a distance, it would be taken as a part of the natural covering of the head.

She looked grave and serious until Eustace turned his head from calling his dog. Then, as she recognized him, her broad, yellow face broke into an expansive smile, as she uttered a pleased exclamation in her native tongue.

It cannot be said that Myeta was handsome,

though she may have had some such claims a score of years before. Her nose was short, but of aquiline form, and when she smiled her little bead-like eyes were almost hidden by the fat and wrinkles. That which was the most displeasing, however, was her teeth, which, though sound and even, had been discolored when she was much younger by some decoction, so that all of them were almost as dark as her hair.

But Hermon had noticed the same peculiarity among the Javanese, and had become, in some degree, used to it. Myeta stood with her arms, naked to the elbows, hanging by her side, and was so pleased to see the son of her husband's employer and good friend that she stepped nimbly from the door with an odd, cackling laugh, and before the amused Hermon saw what she meant to do threw both arms about the neck of Eustace and gave him a hearty kiss upon his cheek.

"She is very fond of you," said the young American; "I don't wonder that you felt you would be welcome——"

At that moment the fat arms of the demonstrative Myeta were flung around Hermon, and he was saluted in the same manner, and with such vigor that his hat fell to the ground.

"That's because you're *my* companion," said the smiling Eustace; "thus we see the advantages of friends at court. She can't talk English at all, though her husband speaks it well; so you needn't be afraid to talk freely before her."

"What's *his* style of receiving strangers?" asked Hermon, carefully replacing his hat.

"Jurak makes great pretensions to being a Caucasian in his manners. He hardly ever speaks Javanese, except where he has to do so to make himself understood. He will shake you by the hand, and show he is glad to have you as his guest."

Having made her salutations, Myeta, with another comical laugh, whirled about and ran chattering into the door, expecting, of course, her visitors to follow her. They did so, Eustace leading the way.

The room which they entered was about a dozen feet square, and was furnished in an inexpensive style. There were three stools made of bamboo, a plain stand or table constructed from Javanese teak, a piece of rag carpet containing three or four square yards, several garments belonging to the wife and husband hanging on pegs fastened in the side of the building, and, with the exception of a few trifling articles, that was about all.

On the stand lay a long-handled Dutch pipe, with a paper of tobacco, from which a lot had been spilled. As seen by Hermon, it looked precisely like dried tea leaves. On another part of the table were a number of cups, saucers, bowls, spoons, and other articles, all of which had been given to Myeta by Mrs. Hadley, the mother of Eustace.

The moment the fat little housewife was inside her home she dropped upon one of the stools, and, folding her hands over her lap, smiled more expansively than ever, and began chattering with Eustace at a bewildering rate. The youth understood the native tongue well enough to sustain a conversation in it, though he could not begin to equal her. Hermon could only sit still and listen and look, without catching the meaning of a syllable. He and his cousin had set their rifles down in one corner, and laid their hats on the floor beside them. Hermon noticed the features of the place which I have named, and his eyes also observed the door which opened into the rear apartment. This was merely an opening just broad enough to permit the passage of a person. A curtain of light-colored cloth was looped at the side, so that it could be dropped down and shut out all view of what lay beyond. Hermon caught a glimpse of the dark floor of earth, but saw nothing else.

Suddenly Myeta, while smiling and chatting at her best, uttered a half-cackling whoop and leaped from her seat as if bitten by a serpent, and plunged headlong out of the door.

"What's the matter? what has alarmed her?" asked the frightened Hermon.

"Nothing; she is impulsive; when she asked me whether I would like something to eat, and I said yes, she set out to get our supper for us. That is her enthusiastic style."

"It is enthusiastic, indeed; but where is her husband?"

"Jurak went up the mountain this afternoon with his gun; he had permission to spend several days home, and to-day news reached him that a tiger had been seen only a few miles away."

"A tiger!" exclaimed Hermon, with a flash of the eye; "I wish *we* could get a chance at him!"

"I don't know about that," replied Eustace, gravely shaking his head; "the tiger in Java is the royal Bengal of India, and you know he is the most terrible creature in the world. He will dash right into a house like this, and tear it into pieces in his eagerness to get at the inmates. They are not very numerous in Java, but we meet them once in a while. This tiger dragged a boy from a cart on this very road, carried him a short distance into the woods, and left only his bones. He is a fearful beast, and if the natives hunt him they will do so in large numbers."

CHAPTER IV.

AN UNWELCOME VISITOR.

"ISN'T Myeta alarmed for her husband's safety?" asked Hermon, a few minutes later.

"No; he has been on tiger-hunts before; therefore he knows the danger, and will be cautious. She says he may be home this morning, and he may be absent for a day or two."

The conversation was interrupted by Myeta, who came through the door as if the tiger was after her; but she was grinning and chattering in the same overwhelming fashion. Removing the pipe, tobacco, and several articles from the table, she picked it up and started out, with the dishes, forks, etc., upon it. Hermon sprang up to offer help, but Eustace told him she did not wish it, and he sat down again.

Within a few moments the boys were asked outside; and taking their stools with them, they sat down in the shade before the table, though the natives, as a rule, sit upon the ground when taking their meals.

Hermon was pleased to note a certain cleanliness, which was not expected from what he first saw. The table itself had been brushed free from dust and foreign particles, and the dishes and everything were devoid of dirt. On the little fire burning near by had been prepared some tea, as cheering and delicious as he had ever tasted. There was some lamb chops also, "done to a turn," and some yellow-colored rice; and the bread, though dark in color, and with a peculiar seedy taste, was light and nourishing. Water, too, was abundant, cool, clear, and sparkling.

The long tramp of the boys had given them keen appetites, and they captivated the impulsive Myeta by the vigor with which they disposed of what she set before them. She did not join them at the table, but narrowly watched the two from a respectful distance, eager to anticipate their every want. They laughed, when, with her queer exclamation, as she saw Hermon transfer the last chop to his plate, she sprang for several uncooked ones, and lost no time in placing them over the fire.

When the meal was finished Tweak came frolicking up to his master, and claimed notice. Myeta understood what he wanted, and the animal was supplied with food. Then, as the cousins sat on the grass outside and chatted over the good things they had had together in America, so many thousand miles away, they noticed that it was night and the full moon was shining overhead.

Hermon had never seen a greater flood of moonlight, even when sailing over the ocean, where he saw many nights of marvelous brightness. The twilight in that latitude is short, and it seemed to him as if it was really lighter than when the sun was dipping below the horizon. The air, while not cold, was delightful, and the long tramp they had taken gave a zest to their idle lolling upon the ground.

Myeta took but a short time to clear away the "things," after completing her own meal. She carried the table into the front room, and saying something to Eustace, passed out of sight into the rear apartment.

Hermon asked what her words meant.

"She told us not to stay out too late, for the tiger might come this way. She bade us good-night, and we shall not see her again until she comes from her room in the morning. I think, Hermon, it will be wise for us to go inside."

The younger cousin laughed at what he looked upon as timidity, but, making no objections, followed Eustace within. They reclined on the piece of carpet, which was drawn up near the door, and looked on the moonlit space between the hut and the road. The illumination was so full that one could have read fine print by it, and the highway was as distinctly visible as at midday.

Hermon was telling his cousin some experience of his when on the steamer which brought him to Java, when he noticed that he made no comment at the conclusion of the story. He asked him what he had to say, and still receiving no reply, turned around to learn the cause of his silence. He saw that he was stretched out at full length, on his face, and sound asleep.

"I'll watch the road awhile," muttered Hermon, with a smile, "and then I'll shut the door and go to sleep."

As may be supposed, less than a quarter of an hour had passed when the drooping head of the American sank to the ground, and he, too, passed off into the land of dreams, leaving the door of the bamboo hut wide open. In such a well-governed country as Java there is little to be feared from marauders, but Jurak and Myeta never thought of retiring to rest without fastening the single outside door, and the guests were expected to do the same; and they would have done it had they not fallen asleep.

When Hermon awoke he found himself lying near the door, with Eustace still soundly asleep, while it was to be presumed that Myeta was in the same condition, since all was silent in the direction of her apartment.

"My gracious! that was careless!" whispered Hermon; "the door has been open all the time we were asleep!"

Not wishing to disturb his companion, he softly rose and moved forward on tiptoe. Nothing had been seen of Tweak since supper, and it was safe to presume he had taken lodging somewhere in the woods.

When Hermon stepped outside he found the moonlight so wonderfully brilliant that he walked forward a few steps and drew out his watch. He saw by the figures that it was a few minutes past midnight.

"I don't suppose they will awake before sunrise——"

Precisely why, he could not tell, but some cause led him to glance to the left, toward the upper part of the highway, in the direction of the Sacred Mountain. As he did so he was transfixed by the most fearful sight on which he had ever gazed. In full view, and no more than a hundred feet distant, on the open space, close to the highway, stood a splendid royal Bengal tiger—one of the largest of his species.

He was evidently coming down the road on a search for food when he caught sight of the young man as he emerged from the bamboo hut and examined the face of his watch by moonlight. The beast, walking as softly as a cat, stopped short, raised his head, and inspected the stranger.

And Hermon Hadley, having advanced several paces from the door, also stopped short, and fixed his eye on the most dreaded quadruped that lives in jungle or forest!

When the youth expressed the hope that he would gain sight of the tiger known to be in the neighborhood, he little dreamed that his wish would be gratified so soon. Certain it is he had no desire to meet the beast under such circumstances, and with the advantages so much on the side of the enemy.

While no animal can be named of which the royal Bengal tiger has any fear, and while he looks upon man as little more than a plaything, it was clear that the one which saw young Hadley was surprised. This was proven by his action, or rather want of action. Within a short time previous he had killed a native youth, and, while prowling around the country in search of others, he suddenly saw a young man come out of a native hut and walk toward him, as if about to offer himself as a sacrifice. Certainly it was not in accord with the rule by which he secured his meals.

But though young Hadley was transfixed by the sight, it was only for a moment. He was quick to realize that it was no time for weakness or hesitation. A hungry tiger "means business" from the first, and it would not take this specimen long to settle on his course of action.

"If I had my gun I would not care," was the thought of Hermon; "but it won't do to stand here."

Instead of whirling about and dashing through the open door of the hut the youth began stepping slowly backward, with his gaze fixed on the ferocious beast. Well aware of the prodigious activity of his foe, he was afraid that if he started to run the animal would be at his heels before he could seize his gun or close the door. The instant the lad showed he was retreating the tiger would leap after him, while a guarded withdrawal might not be noticed until Hermon had secured the advantage he sought.

He was quick to act on this view of the situation, and in his eagerness to lay hand on his loaded rifle he retrograded so fast that the tiger must have discovered what he was doing.

A situation more alarming than that of Hermon cannot be conceived; for only a few steps distant lay his cousin, sunk in peaceful slumber, a most inviting prey to the tiger. Any way, it looked as if the brute was to gain the supper he was seeking, and that, too, without inconvenience to himself.

Hermon had lifted his foot to take the fourth backward step, and he was sure the tiger was gathering his mighty muscles for a spring toward him, when some slight noise in the wood behind the animal arrested his attention. The youth saw the head flirt backward, and then, as noiselessly as a shadow, the animal turned, so that he stood broadside toward him, with his inquiring gaze centered on the suspicious point in the rear.

This was Hermon's chance. Turning like a flash, he made a couple of quick bounds, and was within the bamboo hut. The bright moonlight streamed through the open door and fell on the feet and legs of Eustace, the upper part of whose body lay in shadow. Remembering just where the guns stood, Hermon caught up his own and stepped to the front.

The tiger had vanished. Not the slightest noise betrayed his departure, but the keen vision of the youth, roving over the brightly-lit opening, saw nothing of him.

And yet, for all that, he might be closer than before. Hermon shuddered as he reflected that possibly he was in the act of stealing upon him that very moment. He lost no time in closing the door and putting in place the ponderous wooden bolt with which Jurak had provided it.

"What are you doing?" asked his cousin, only half-awake.

"Only fastening the door against that tiger which is outside," replied Hermon, grimly enjoying the consternation he knew his words would cause.

"What!" gasped Eustace, coming to a sitting position with the suddenness of a jumping-jack.

"I saw that tiger—and he is a splendid-looking creature—standing outside in the opening, and I think he wants to come in here!"

"If he wants to come in, nothing will keep him out!" exclaimed Eustace, who was on his feet, gun in hand, the next instant. "I tell you," he added, after glancing out of the narrow window without catching sight of him, "that the tiger is the worst animal in the world. Old hunters will say that. Sometimes five hundred men will engage in hunting a single beast, and every one of the five hundred will tremble in his footsteps when he knows he is anywhere near the creature."

"I supposed the real royal Bengal tiger was found only in India."

"Java was once a part of the continent of Asia," said Eustace, giving expression to the generally accepted theory about that remarkable country, "and we have a good many of their animals."

While the two were speaking, one was standing at the narrow window on the right and the other one on the left. Though separated by the breadth of the room, it will be understood they were quite close together. They spoke in low tones, naturally timid about attracting the notice of the fierce beast outside, and desirous at the same time of not disturbing the slumber of Myeta in the next apartment. Hermon had given his account of his thrilling experience, and Eustace shuddered to think by what a narrow chance the whole three had escaped for the time.

A few moments later, the tiger, advancing toward the open door, would have pounced upon the sleepers and made short work of them. It was providential indeed that Hermon awoke when he did.

"Do you think he will come back?" asked the younger.

"I am afraid so; I hope not, for I would rather see any wild beast in Java than him; but, if he does come, I think it likely he won't molest us, if we don't molest him. You see he will be a little puzzled by the hut, and he will walk around it, studying the best way of attacking it. A tiger likes to see some signs of life before he leaps upon his prey, and if all is still within he may go away without assailing us."

"Then, if he lets us alone, you think it best for us to let him alone?" was the inquiring remark of Hermon.

"Yes, indeed! If he shows himself, don't stir or make the least noise."

"That depends on circumstances," was the mental remark of Hermon, who could not bring himself to feel the dread of the tiger which showed itself in every word of his cousin.

Perhaps, if he had been as familiar with the animal as was Eustace, Hermon would have shared that dread with him.

CHAPTER V.

A RIFLE-SHOT.

THE COUSINS exchanged a few more words and then became silent. They felt it was hardly the time for conversation, and there was need of being vigilant.

Eustace was hopeful that the disturbing cause which had drawn away the attention of the tiger at the critical moment had resulted in his departure from the vicinity. Knowing the amazing strength, courage and ferocity of the brute, the youth would be timid about venturing from the hut, even during the day-time.

Hermon did not know whether he wished the brute to go away or not. Sometimes he thought that if Eustace was so fearful of him he ought to feel the same; but there was a lingering natural desire that he might be given the chance for one good shot.

"*By George! there he is!*" suddenly whispered Eustace.

Hermon stepped lightly forward and peeped over the shoulder of his cousin, who said, in the same guarded voice:

"He's just gone around the rear; he's reconnoitering and trying to make up his mind which is the best way of attacking us. Be careful!"

Hermon was back to his own window just in time to gain full sight of the magnificent brute, which, as Eustace had said, was doubtless studying the best way of assailing the hut. As he walked around the suspicious object with his cat-like tread, he was less than fifteen feet distant. He was a royal creature indeed, and stood revealed in the moonlight the very embodiment of grace, beauty, and tremendous muscular power. No wonder that he feared nothing, and looked upon every living creature as his prey.

When he came directly opposite Hermon, the latter raised his rifle, and with the muzzle projecting a few inches from the narrow window of the hut took dead aim at him. Eustace, who was looking for the brute to appear each moment on his side, did not notice the slight click of the gun-lock, and therefore had no suspicion of the purpose of his companion.

With a coolness that was surprising, Hermon waited until the tiger thrust forward his fore leg, so that the route to his heart was left open, and then he fired.

It would be hard to say which was the more amazed for the instant, the tiger or Eustace Hadley, for the shot was as unexpected to the one as to the other.

"Gracious! what have you done?" exclaimed Eustace, bounding to the side of his friend.

"I've killed the tiger," replied Hermon, bending forward and looking out of the window.

Astonishing as was the declaration, it was the truth. Everything united to help Hermon. The distance separating him from the brute was slight, and he was given the opportunity to make his aim certain. He held one of the best of weapons in his hand, and Gordon Cumming himself never fired with more coolness.

No matter how enormous the strength and vitality, no matter how fierce the animal, one little ounce of lead directed to the right spot will bring it low. The tiny bullet through the heart of the behemoth drives out life just as surely as does the arrow which cleaves the heart of the sparrow on the wing.

When hit by the rifle-ball the tiger made one terrific bound straight upward, as though an immense spring-board had flung him aloft, and, landing on his feet, he rolled over, clawed the air viciously, snarling and writhing in pain, and then he ceased his struggles and became still; he was stone dead.

The report of the gun startled Myeta from her sleep; and, thrusting her head through the curtained door, she asked in a frightened undertone what it meant. When Eustace told her that the rifle that had just gone off had killed the tiger, she could not believe him until she had hastily dressed and come forth.

"That's the most wonderful thing I ever knew," said Eustace; "it seems incredible, and I can scarcely realize it; it will make you a hero among all who hear of it."

"I don't understand why it should," replied Hermon. "I can see why it is considered an exploit for a hunter to shoot a full-grown tiger, but with me it was a piece of good luck. Had we met him in the woods, it isn't likely anything of the kind would have taken place; but he came up within the right distance, and a poorer marksman than I could have sent a bullet through his heart."

"By George!" added Eustace a minute later, while unfastening the door, "if you had fired on him when he stood with his head toward you, out there in the opening, there would have been a different story to tell. The whole thing was providential, indeed! Let's take a look at him!"

"Isn't there any danger from his mate?"

"No; they don't often travel together; but we may as well take our guns with us."

The youths, each with his rifle in hand, and accompanied by Myeta, as much interested as they, passed through the open door of the hut and over the brief space to where the dead monarch of the jungle lay.

With a lingering fear that he might not be dead

they kept a respectful distance for a minute or two; but it was soon apparent that not a spark of life remained. Myeta did not approach close, but the boys pushed him with their guns, and even kicked the limbs and powerless legs.

Ah, if the beast had been alive when that indignity had been put upon him!

One suggestive fact was noticed by Hermon and Eustace; the sharp claws and teeth of the tiger showed that he had recently eaten food. They were covered with the evidence, and it was more than likely that the slight noise which distracted his notice when Hermon was stealthily withdrawing from before him saved the life of the lad, but it did so at the expense of another life, which was thus involuntarily given for that of him who, in the end, slew the tiger himself.

The little party did not spend much time in admiring the dead tiger, for the hour was late, and Hermon could not free himself from a fear that another one of the animals was prowling in the vicinity. They returned to the bamboo hut, therefore, and fastened the door with unusual care. Myeta withdrew to her apartment, and the boys, stretching out on the carpet, soon fell into a sound slumber which lasted until sunrise, when it was broken by their hostess.

Hermon was hopeful that Jurak would reach home before their departure, but he did not. Bidding their kind hostess good-by, after a morning meal, they left, following the well-traveled road which ran almost due south. The elevation steadily increased, and a gradual change in the temperature was felt. The dog, Tweak, did not show up, and they therefore went away without him.

The sky was clear, and a gentle breeze fanned their faces, which became somewhat flushed from the hard exercise of climbing what may be called the backbone of the land.

"You see," said Eustace, "there isn't any doubt that Java, like many other islands around us, was once a part of the continent of Asia."

"What proof is there of that?"

"There are lots of them; the volcanoes themselves show that great upheavals have taken place. I am speaking, you understand, of Borneo and Sumatra, as well as of this island. In those two I have named you will find the elephant and tapir, and in Sumatra and this island are the rhinoceros. Many of the wild cattle which they used to believe were peculiar to Java are now known to exist in different parts of Asia."

"There was no way by which they could have been brought across the water, I suppose?"

"The only way I know of is by ships and vessels, which, I believe, those animals have never been suspected of building."

"How about the birds and insects?"

"They add to the proof. Some people naturally think that birds will pass from one country to another when the water separating them is not too extensive, but it isn't so. In Java we have many kinds of birds which never cross over to Sumatra, only fifteen miles away, with islands in the middle of the strait that separates us. We have more birds and insects peculiar to ourselves than have either Sumatra or Borneo."

"What does *that* show ?"

"That Java was detached from the continent long before Borneo and Sumatra. Another fact is that the sea between these islands and Asia is shallow; but," added Eustace with a laugh, "I don't see the need of my going into such an argument, when I haven't heard you dispute it."

"No, I am satisfied that not only these islands, but the Philippine, and all that string off eastern Asia, once belonged to the continent, and in some vast upheaval they moved out to sea, and will stay there until another overturning comes."

The road which the youths were following still wound through the woods, among the rocks, and across patches of open land, where the grass grew luxuriantly on both sides of the highway. A short distance from the hut of Jurak they met a couple of natives. They were fine specimens of young Javanese, though when Hermon looked upon them he felt that it would be an easy matter to fling each over his head in a wrestling bout. They lacked the strong, lusty vigor and admirable proportions that are shown to perfection in the Caucasian race.

In fact, these two youths were excellent examples of the Javanese branch of the Malay race, but there was nothing attractive in their appearance. Their narrow, sloping shoulders and their slight figures were not likely to inspire respect or fear. Their faces were irregular in shape, and the eyes of one were raised at the corners, as is often seen among the Mongolians. Their noses, which were not flat, though short, saved their features from real ugliness. The lips of both were thick and projecting, and on the upper ones were a dozen or so puny hairs—the nearest approach to a mustache. However, they were quite intelligent, and the fact that each carried a long spear suggested to the boys the errand on which they were engaged.

A short but spirited talk with Eustace in the Javanese tongue made known the general consternation that the tiger had caused. He was known to have slain several people, and I may as well add, what I have hinted, that the slight sound which drew the creature away from his attack on Hermon Hadley, when he stepped out of the hut, was caused by the approach of a person who never knew his danger until the tiger bounded upon and bore him to the earth. The bones of this poor fellow were afterward found within twenty rods of Jurak's hut.

The beast could not have been famishing directly after, but his taste for blood was not fully satisfied, as was shown by his prowling around the place in which the youth had taken refuge.

The inquiries made by Eustace showed there was but the single tiger that caused all this alarm. In the luxuriant forests beyond them, stretching for many miles toward the rocky southern coast, were found the black as well as the Bengal tiger, besides numerous other dangerous wild beasts. So long as the worst of them kept within those limits, little harm resulted; but now and then, as in this instance, some animal would come outside and create havoc and a general panic.

The tiger had been heard a number of miles away, and was known to be working to the northward. The natives, whom the boys met, were out to summon several hundred of the neighbors for the purpose of hunting the animal. Their method was to surround him in some jungle, swamp or thicket, and then gradually narrow the circle, until he would rush out and attack those nearest him. The moment he did so, a multitude of spears would be buried in his body.

When Eustace Hadley told these young men that the beast was shot the night before by the modest-looking young gentleman standing a few feet away and listening to their words, without understanding a syllable, they hardly believed him; but he insisted, and gave the particulars until they could not refuse to credit the amazing story. They surveyed Hermon with such admiring looks that he flushed and was embarrassed.

"Eustace, what are you telling them about me?" he asked.

"When a fellow comes all the way from America and kills one of the biggest tigers in Java, why, you don't suppose the story is going to shrink while I have charge of it? Not if I know myself," said Eustace, laughing and enjoying the confusion of his friend. "After I had made them believe you really killed the animal, they were ready to swallow anything I might say as to the methods. What if I did relate how you sprang from the limb of a tree and landed on the back of the tiger; that you held him by the ears, so he could not turn to bite you, and then wrestled and finally flung him off the rocks."

"Oh, my gracious!" exclaimed the disgusted Hermon, turning about and starting up the road; "I can't stand everything!"

Eustace laughed merrily. Then he hastened after his cousin, calling out that he had told the youths nothing but the simple truth, for that was enough to cause their eyes to protrude with wonder.

When Hermon stopped, after walking a short distance, and looked around, he observed the two Javanese still standing in the middle of the highway and staring at him, as though he was the greatest hero that had ever set foot on the island. He began to fear that this business was likely to become anything but pleasant to him.

The boys had gone less than half a mile when they encountered three other natives, each of whom held a long, formidable spear. Hermon begged his cousin to make no reference to him in whatever was said to them, but the conversation hadn't lasted two minutes when the three turned about and gazed with such fixedness and admiration on the young American that he knew Eustace was giving the story again in his most flowery fashion.

"I'm getting tired of this," he said. "How many more of these simpletons are we likely to meet?"

"Well, you know, Java is one of the most thickly-populated countries in the world, and this part of the island seems to be up and in arms on account of the promenade of that single tiger. Oh, I don't suppose," said Eustace airily, "that we are likely to meet more than several hundred more this forenoon!"

CHAPTER VI.

THE LONELY LAKE AND A STARTLING ADVENTURE.

A SHORT distance farther, and Eustace, acting as the guide, turned abruptly to the left, and entered the wild, broken country where there was no highway, and the only paths were such as were worn by the feet of wild animals. The Indian Ocean, with its rocky, volcanic shores, was scarcely twenty-five miles in a direct line to the south, while the low, unhealthy marshes that stretch along the line of the Java Sea were still less remote to the north.

The island, as I have shown, is one of the most populous in the world, and is so well governed that nothing is to be feared from lawless men.

In the depths of the mountainous region, however, on which they were entering, were many strange and fierce wild beasts. The youths would have been reckless beyond excuse had they attempted such an exploration without their fire-arms.

Hermon noticed that for a time they ascended no higher. Instead of climbing the mountains, some of which still towered far above them, they seemed to have reached the borders of a large plateau, where rocks, forest trees and vegetation were so abundant that they could see only the briefest possible distance in advance. Had they not made full preparation for this sort of traveling they would have become lost within the space of half an hour. Each carried an excellent pocket-compass, and as the elder was familiar with the contour of the island and the precise direction they must take in order to reach any point, they gave little heed to their footsteps, but wandered along like a couple of truant children, careless whither their steps led them.

And during these wanderings it must not be supposed that they saw but few signs of animal life. They were hardly ever out of sight; and many an exclamation of amazement and delight fell from the lips of Hermon, who for the first time in his life met numerous wonders of the tropical forest. They had not yet passed the boundaries of the second zone in Java, which I have described elsewhere. In the first were the rice-fields, sugar plantations, cocoa-nuts, cinnamon and cotton, while in the one which our young friends were traversing were the coffee and tea plantations, the sugar-palm and the maize.

The cousins were pushing their way in this leisurely fashion when Eustace, who was in advance, brought his gun to his shoulder and fired.

"What is it?" asked Hermon, unable to see clearly on account of the intervening vegetation.

"Something for dinner—a kidang or mintjac."

"I am glad to know it," said Hermon, who had never before heard the queer names; "I thought it was that sort of bird. Did you take him on the wing?"

Eustace laughed, and after advancing a few steps paused beside a delicate little deer, which he had shot. He explained that the kidang and rusa are two kinds of deer that are hunted in Java for sport, and sometimes for their food, both being excellent. When a portion of the creature was dressed and cooked over a small fire it formed a nourishing and palatable dinner, seasoned as it was with the mixture of salt and pepper with which each was furnished.

"If I am not mistaken," said Eustace, looking around as though he recognized the spot, "we are not far from a small lake."

"I feel thirsty," remarked Hermon, "and was wondering how much longer you would go before suggesting that a drink wouldn't be a bad thing."

"There are plenty of pitchers to carry it in," added Eustace, with a smile which brought a nod from Hermon.

"I have noticed a good many of them ever since we left the highway."

The boys were referring to one of the many strange vegetable productions found in the tropical regions—the pitcher-plant. It grows where the soil is damp, and in the moist climate of Java flourishes vigorously. It has narrow, cylindrical pitchers, some six or eight inches in length, and of the same light green color as the leaves. The pitcher-plant would not be so remarkable if the mouth was not furnished with a perfect lid, attached by a sort of hinge, and is sometimes open and sometimes closed. The lid does not open until the leaf is entirely developed. Previous to this the watery liquid is secreted, and partly fills the pitcher. The belief generally prevailed only a few years since that the leaves secreted fluid in this fashion for the use of travelers in arid regions; but such is not the case, as the plant never grows in a dry atmosphere.

Hermon plucked a number of the curious pitcher-plants and tasted the water within. He would have been glad to consider it cool and delicious, but he could not bring himself to swallow any of it. There was just enough vegetable flavor to spoil it for such purposes. He preferred to wait until they came upon the fluid in its natural form.

"I thought so!" called out Eustace, as the silvery gleam of a sheet of water was discerned among the trees ahead. "I was here two years ago with Jurak, and was sure I could find the lake again."

The "lake," as he termed it, was of irregular contour, and covered perhaps ten or more acres. The shores on every side were composed of dark bowlders and masses of rocks, all of which were of volcanic formation. Hermon did not need to be told that he was looking upon the crater of an extinct volcano—

one which, after vomiting forth its millions of tons of mud, perhaps, and ashes, had subsided and remained dead for centuries. Gradually the immense scar left in the side of the earth had been filled during the rainy season, and, encouraged by the moisture without, innumerable little springs had begun bubbling through the flinty floor, until a lake was formed with water mildly cool and quite clear.

When the young American looked upon the surface it was without a wrinkle. If there was any air astir, it did not reach this lonely lake.

"Are there any fish in it?" asked Hermon, after they had viewed the body of water for several minutes.

"Hundreds and thousands of them. If you will watch the water closely, you will notice them swimming about."

"Where did they come from?"

"You can answer that as well as I. I suppose there must be some underground communication with the sea through which the fish have found their way to this spot. They are quite small, so far as I can judge, and are not worth fishing for."

The pool looked so charming that the boys stood a long time on the rocky shore, their eyes wandering over the mirror-like surface. In the distance they observed several birds of brilliant plumage flitting back and forth, but they were too far away to be identified.

"Let us make a circuit of the lake," said Hermon, stepping off to the right in the hope of reaching ground that was easier to traverse than the bowlder-like banks.

"We shall find the other side very much the same as this. I went around it with Jurak. But we have enough time to do whatever we choose, so come on."

"Halloo! here is the path of wild beasts," called out Hermon, a rod or two ahead of his friend.

The path to which he referred resembled some of the buffalo trails seen in the West. It was less than three feet wide, and six or eight inches deep; where the ground was soft it was still deeper.

Eustace identified the trail at once.

"That's good!" exclaimed Hermon. "I'm going for big game only—such as tigers, rhinoceroses, leopards, buffaloes, etc. The next one in order is a rhinoceros."

"The animal isn't half as dangerous as the tiger, I'll admit; but because a piece of good fortune gave you the king of the forest, you mustn't rate the rhinoceros too lightly. He is an ugly customer, and I don't know that I ever heard of one being killed by a single bullet. For that matter, the tiger generally takes more, I believe."

"It looks to me as though this path leads right around the lake. I wonder whether a lot of rhinoceroses, when they feel frisky, are not in the habit of trotting around the water playing tag, just for the fun of the thing?"

"The creature is too bulky to dance about in that fashion; the path has been worn here for years. Jurak spoke of it as having been made long before we saw it. Probably the ancestors of these animals were accustomed to take a regular walk to the pool for their supply of water."

"Ah! the path makes a turn here and leads to the south—that is, toward that high mountain chain."

"It passes over that to the dense forest on the other side, where the creatures spend a good deal of their time."

"My recollection of the rhinoceros is that he lives chiefly in the lowlands."

"That may be the case; but in Java our animals don't copy after others, but have ways of their own. In Africa, you know, the rhinoceros has two horns, but those of India and Java have but one. Besides that, our animal isn't so bulky and awkward as the native of India; he stands higher, and runs faster."

"It is a wonder to me how such a hogshead on pegs can run at all," said Hermon with a laugh.

"If you start in a race with one of them you will quickly see how they can run. Gordon Cumming says a horse and rider can rarely overtake one of them. His senses of hearing and smell are wonderfully acute, and he is so big that he isn't afraid of anything. An elephant has been killed by a rhinoceros, and if one of them could get a fair lunge at such a tiger as you shot last night it would be enough to finish him. But, halloo!"

A loud, whiffing grunt or snort startled the lads at that moment, as they were moving slowly along the furrowed path, Hermon still in advance. While discussing the interesting facts about the extraordinary beast they were brought to an abrupt stand-still by the noise described.

A few rods ahead, in the path, stood a striking specimen of the Javanese rhinoceros—such as was partly described by Eustace Hadley. He was not more than a hundred feet distant, and, having scented the approach of strangers, had stopped short, with his ugly head thrust forward, so that it extended in a straight line from his ponderous shoulders. He formed a curious and repellant figure that would check the most veteran hunter.

"We must run!" said the elder youth.

"It seems to me there isn't much use of running when he is so fleet," replied Hermon, keeping his eye on the frightful beast and stealthily raising the hammer of his rifle. "He isn't quick with his eyes, and we have a good chance to dodge him among these trees; or we can climb one of them. Halloo! here comes the old fellow!"

The rhinoceros seemed to have inspected the lads long enough to learn they were intruders, when, with another whiffing snort, he made for them!

"Come, Hermon, it won't do to wait a second!" called out Eustace, wheeling about and dashing off at full speed. Having warned his cousin, he supposed he would lose no time in following him. Both were fleet of foot, and the intention of the elder was to flee until he caught sight of an inviting tree in which both could find refuge.

But Eustace had taken only a few steps when he was startled by the report of Hermon's rifle. Turning, he saw the sturdy young American standing in the path, and in the act of lowering his weapon, which he had just discharged at the advancing rhinoceros. The latter was charging at a speed that was highly dangerous.

"Run, Hermon! Run, or you'll be too late!" shouted Eustace, terrified beyond measure by the peril of his cousin.

The latter could not fail to see his situation. At the moment Eustace called to him he whirled and ran. Instead, however, of following the trail, as his cousin had done, he darted off among the trees, which were so open that the rhinoceros was not visibly impeded in the pursuit.

Inasmuch as Hermon was a good shot, and the intervening space was so slight, it is probable that the bullet struck the beast; but if so, it inflicted no more harm than if it had flattened itself against the face of a rock.

Recognizing the point, too, whence came the

hostile shot, the rhinoceros promptly turned off from the path in hot pursuit of Hermon Hadley, who so underestimated the speed of his pursuer that, while casting about for refuge, he suddenly found the rhinoceros so close upon him that he had no time to escape except by climbing the nearest tree which presented itself.

The youth bounded forward; but, do his utmost, he could not equal the swiftness of his ponderous enemy, which would have trampled him to death the next moment had not Eustace dashed across their line of flight and discharged his gun almost in the face of the pursuer.

The latter did not stop, but flung up his head with an angry sniff, as though he was "struck hard," and, slackening his gait, looked around to learn whence came the shot. Eustace, confident that he would make for him, wheeled and bounded for the nearest tree. In his effort the stock of his rifle caught in a bush and was flung a dozen feet from his grasp; but he had no time to tarry, and climbed with more speed than ever before, never looking below until he was safe among the limbs.

Then, when he did so, he found he had made a great mistake, for the rhinoceros had not followed him at all. After turning his head, as if to identify his second enemy, he continued his pursuit of Hermon.

The slight interruption gave the younger hunter

the few seconds that saved him. He saw he had no time to select his refuge, but must take the very first that presented itself. As his rifle would be only an impediment in that critical moment, he flung it aside and scrambled up the nearest tree with the same haste that his cousin was putting forth at that moment.

In the case of Hermon, he was not a second too soon. He really believed if he had not been very spry in drawing up his legs at the instant he got his arms among the limbs they would have been crushed by the lunge of the maddened rhinoceros, who could reach higher than would be supposed. The fugitive lost no time in climbing beyond the reach of his enemy.

But Hermon was far from feeling safe, for he was sure that two or three shocks, such as came from the charge of the beast, must overturn the tree. Indeed, he thought it was going over as it was. Certainly its roots were loosened, and had the animal possessed enough intelligence he could have dislodged his intended victim by continuing his method of attack.

But the rhinoceros is not bright, and, having made one savage lunge, he stepped back and seemed to inspect the frightened youth above him. As the beast retrograded a brief distance he stepped upon the lock of Hermon's rifle and moved to one side, as though it caused some discomfort. Bending his head, he began rooting it with his single horn, as if he felt a curiosity to learn its nature.

The tree in which Eustace had perched himself was not more than twenty yards distant—a space so slight that, despite the abundant vegetation, the friends were in sight of each other. Feeling quite sure of their safety, Eustace called across, when the rhinoceros began rooting the gun:

"Look out, Hermon! He is going to shoot you!"

"It does look that way. I'll be glad to have him fool with the gun if he will only let this tree alone. If he bangs his head against it as he did a minute ago, that will be the end of me!"

"Your tree isn't as large as mine, but he doesn't know enough to push it over, even if he is strong enough."

"Now, if we only had our guns, we could just sit here and bombard him until he surrendered or our ammunition gave out!"

"It would be a good idea, but neither of us had any time to take our rifles."

"I don't see why you didn't have all the chance you needed."

"I suppose I did, but, as I didn't think so, it was all the same as if I hadn't a second to throw away."

"Isn't there any way of getting our guns?" asked Hermon, peering carefully downward at the brute beneath.

"I have been asking myself the same question.

It seems to me that one of us ought to be able to slip down and get hold of a weapon without being seen by the rhinoceros."

"You tell me he can't see very well, so what's to hinder?"

"I believe he would hear us if we walked in our stocking feet."

"Well, when one starts, the other will make all the noise he can to divert his attention. If you'll open on him with your revolver, I'll try it."

"No," said Eustace; "I fired at him a few minutes ago, but he stuck to you. He is more anxious to have you than he is to get me. Draw your pistol and begin popping away at him, while I sneak down after my gun."

The indications were in favor of this plan, for Eustace's rifle lay nearer to his perch than Hermon's did to his place of refuge, while the enemy was close in under the younger, and acted as if he meant to stay there until the lad should be compelled to descend.

In truth, the course of the rhinoceros was singular. He appeared to be roused to a revengeful feeling toward Hermon, who fired at him when he was advancing, and thus far he paid no heed to the other young hunter, who sent the second shot. There was reason to believe the latter did some execution, while the former did not.

The rhinoceros having lurched against the trunk

of the tree, which was no more than eight inches in diameter, yawed, as may be said, until he stepped on the rifle. He nosed that a minute or so, but soon withdrew his attention, and, stepping back several paces more, became stationary.

"He may grow tired of waiting and go away before long," suggested Hermon, who watched with considerable misgiving the preparation of his cousin to descend the tree.

"I think he will become tired much sooner if I can get hold of my gun," replied Eustace, keeping his eye on the brute. "Are you ready?"

"Yes, if you are determined to try it."

"Of course I am. Pop away, and see whether you can lodge a ball or two in his eyes."

CHAPTER VII.

A TREACHEROUS REFUGE.

"I DON'T see why I can't do something with my pistol," thought Hermon Hadley, as he drew his revolver from his pocket and leveled it at the huge snout of the rhinoceros. "He isn't far away, and if I can send the bullet to the right spot it ought to be equal almost to a rifle-ball."

The youth was astride a limb some twenty feet from the ground, and his target was so large that it would seem he ought not to miss; still he would have preferred to have the beast a little closer. A large branch interfered with his aim, and he broke it off and flung it aside. His cousin was in his field of vision, so that he saw every movement made by his friend.

Sighting the revolver as best he could at the right eye, Hermon fired two cartridges in quick succession. He was sure that one of them struck and glanced off the single horn of the rhinoceros, while, so far as could be seen, the other missed him altogether.

Eustace, having settled on his course of action, did not hesitate. He was not very high, and it was an easy matter to descend. Hermon could not restrain a smile when he observed him sinking on the opposite side of the tree like a person timidly letting himself down into a well. Glancing from the rhinoceros directly toward his friend, he could see his knees and hands as they clasped the trunk, while he peeped first around one side and then around the other; but he steadily sank lower and lower, until his feet touched the ground.

When Eustace stepped from behind the tree and began stealing his way toward his gun, lying midway between him and the rhinoceros, the situation became delicate and critical.

"I don't think he will discover him," thought Hermon; "but there is no telling which way he will turn, and his ears are sharp enough almost to hear a pin fall."

The rhinoceros was an odd sight as he stood with his immense and ungainly head turned toward the lad perched in the tree. He did not seem to be watching him, but held his position motionless, with his blinking eyes fixed upon nothing at all. Now and then Hermon discharged a cartridge; but he might as well have saved his ammunition, for he not only failed to injure him, but did not seem to draw any additional attention to himself by these demonstrations.

The position of the gun which Eustace was so anxious to secure was not as favorable as he could wish;

but he felt hopeful of success, for the distance was slight, and at such a time a person is sure to possess much activity.

Step by step he advanced on tiptoe, his eyes fixed on the enormous creature, while, half-stooping, he held his hand extended, ready to grasp the rifle the moment he was near enough.

When only three steps separated Hermon from his prize he stooped still lower and moved with greater caution than before. He had gone so far that he was certain of obtaining the weapon, for even if the beast turned upon him he meant to seize it before fleeing.

The rhinoceros, from some cause or other, did move his head, and caught sight of the crouching figure. Instantly he charged toward him.

"Quick!" shouted the alarmed Hermon, who let fly, with the remaining charges in his pistol, at the side and hind-quarters of the beast.

But Eustace had observed the preliminary movement of the rhinoceros, and lost not a second. His right hand shot forward, and snatching up his gun he darted among the trees, determined to cling fast to the weapon; but he was disappointed after all. The rifle was not so valuable as his own life, and before he had run twenty steps he dropped it and leaped into the branches of another tree, being barely able to scramble up in time to save himself.

"My gracious!" he gasped, looking down at the

brute. "That campaign of mine wasn't much of a success."

He was somewhat further removed from Hermon, but by leaning forward and parting the branches he was able to catch sight of his companion.

"Hermon," said he, "I made a pretty good run for it, but he was too fast for me. What did you think of the race?"

To his astonishment Hermon made no answer. He saw him sitting on the limb, holding on with both hands and seemingly looking at something on the ground. Eustace called several times, but was alarmed to receive no reply. Hermon seemed stricken with deafness.

"What can be the matter?" thought the elder, forgetting for the moment the brute in his anxiety for his friend.

He had hardly asked himself the question when, to his horror, Hermon suddenly rolled to one side and tumbled to the ground, striking on his side, where he lay as if dead.

The heart of Eustace almost stopped beating. He sat staring at the motionless figure, and was on the point of leaping from his perch and running to his help. But he knew that if he did so he not only would not help matters, but would assuredly be trampled to death himself.

It seemed remarkable that the rhinoceros had not seen the fall of the other; but for the present he acted as though there was but a single person in the world, and he was the enemy in the tree just above him.

Eustace was not only terrified but dumfounded by the extraordinary accident that had befallen his companion. He was unable to account for it. Had they been in a wild country, he would have believed he had been pierced by the poisoned arrow of some concealed savage in the bush; but that was impossible, though some of the Javanese use such weapons in hunting certain species of game.

Gradually the hope took shape that his friend was not dead, but had been smitten by some sudden illness, which would soon pass away. When he detected a slight movement of one of his feet Eustace was relieved, but quickly gave way to alarm as he reflected again upon the danger to which Hermon was exposed from the rhinoceros. The former, however, was quickly reminded of his own peril by the action of the ponderous brute. Ramming his head against the tree, he began rooting it with his horn. He tore off the bark and gashed the wood so vigorously that he was sure to bring the boy to the ground if he continued his work only a few minutes longer.

But the rhinoceros kept up his reputation for doing unexpected things by abruptly ceasing when it may be said he was on the point of bagging his game. He snuffed the air as though he scented danger, and then, swinging his fore-quarters around, moved off at an easy pace toward the deeply-worn trail from which he had ventured in his pursuit of the boys.

This course took him so close to where Hermon lay that Eustace held his breath, fearing that he was aiming for the lad; but he lumbered on, and a few minutes later vanished from sight.

Without waiting, Eustace caught up his gun and ran to his prostrate friend. At the moment he reached him, Hermon sat up and looked around in a dazed way. His face, generally full of color, was as white as a sheet.

"You are sick!" exclaimed Eustace.

"Sick!" feebly repeated Hermon. "I was never so ill in my life! Can you get me a swallow of water?"

Eustace ran in the direction of the lake, breaking off a pitcher-plant on the way. He emptied this of its fluid, carefully filled it with water, and hastened back to Hermon, who had managed to get upon his feet and walk a few paces, when he was forced to sit down on a bowlder of sandstone.

He took several swallows of the cool fluid, and immediately felt better. He was something like his bright self when he smiled, looked around, and asked:

"What has become of that rhinoceros?"

"He is gone. I don't think there is anything

more to fear from him. But tell me what this means, Hermon?"

"That's what I should like to know. It's the strangest thing I ever heard of. I broke a limb in front of me, so as to gain a better view of you while shooting. Just after that I noticed the most sickening odor you can imagine. I never knew anything like it, and the hand with which I broke off the branch was inflamed and itched terribly. It is nearly all gone now, but you can see it yet," he added, holding up his left hand, which was red and pimpled, as though violently poisoned.

"I couldn't tell what to make of it," he continued. "I hoped I would be able to overcome it, but the first thing I knew my head began to swim, and just as I heard your words in my ears, without being able to reply, over I went. If the rhinoceros had been below at that time he would have had me sure. What's the matter?"

Eustace Hadley ran a few intervening paces, and made a critical examination of the tree from which his cousin had fallen. The trunk was less than eight inches in diameter at the largest part (as has been stated before), and there were no limbs until a point about ten feet from the ground was reached. Then they branched out regularly, tapering symmetrically toward the top, so that the upper portion of the tree was of the shape of a bee-hive. It was covered with green leaves, and the piece of branch which Hermon had flung to the ground exuded a resinous sap whose odor Eustace noticed before picking it up. He gave it a slight sniff, and then, with a gesture of disgust, flung it from him.

"Hermon," said he, turning about and walking toward his cousin, "I understand why you became so ill."

"Why was it?"

"That," pointing to the limbs from which the other had fallen, "*is the upas-tree!*"

"What!" exclaimed Hermon, forgetting his weakness. "Let us hurry away!"

"What for?" asked the elder, with a smile, though he took care to walk a few yards further off. "It is poisonous, but not so poisonous as you think. If it was, where would you be?"

Reassured by the words of Eustace, Hermon sat down nearer the shore of the lake. He was soon almost as well as before; and, having recovered his gun and found it unharmed, he reloaded it, and felt as though he would like to have the rhinoceros return.

"The upas-tree!" he repeated wonderingly to himself. "Who has not heard of that? I remember reading in one of our school-books an account of the upas-tree of Java which fairly made my hair rise."

"That all came from a surgeon named Foersch, who was in the employ of the Dutch East India

Company nearly a hundred years ago. He said it destroyed life within a radius of ten or twelve miles, and that out of ten persons who approached it only one lived to return. We have his account at home, where he tells how many minutes a fowl, a dog, and other creatures, lived after being lowered in the valley where one of the trees was growing."

"But he surely had cause for his stories?"

"Yes; that can't be denied; but in 1810 Leschenault disproved the statements of the Dutch surgeon. You see, the upas grows among other trees. Lizards and wild animals do not avoid it, but there is no use of denying that it is very poisonous. The Javanese have used the exudations for years with which to poison their arrows. Had you not broken off that branch, so as to come in contact with the juice, and so as to get the odor directly in your nostrils, you would never have known you escaped from a rhinoceros by climbing an upas-tree."

"I think I have read that it belongs to the breadfruit family."

"Yes, and botanists now unite it with the mulberry family. Like a great many wonderful things of which you hear and read, it doesn't prove near as wonderful when you come to investigate it."

"There's one thing quite certain," said Hermon with a laugh. "I don't care how close a rhinoceros is upon my heels, I'll keep on running till I find something better than a upas-tree for a refuge.

Don't you remember the lines of Fitz-Greene Halleck, in his address to the Indian Chief, Red Jacket?

> That in thy veins there flows a fountain
> Deadlier than that which bathes the upas-tree!"

CHAPTER VIII.

IN THE JAVAN FOREST.

A SOFT rustling of the leaves caused the boys to turn their heads. Tweak, their dog, came walking forth with head erect and gazing right and left, as if to say he was very sorry his engagements had kept him away when they were in such danger from the rhinoceros. But for that, he would have driven off the beast.

"I believe Tweak is a coward," said Hermon. "He never showed himself when the tiger was prowling around Jurak's hut; and now, when the rhinoceros is beyond sight, he comes out of the woods."

"Is that dog cowardly or wise which keeps out of the way of a tiger or rhinoceros?"

"Wise, if he understood the situation; but a brave dog doesn't stop to count the cost when danger threatens his master. If he had shown himself, when the rhinoceros had us treed, as a guaranty of his good faith, I would have had a much better opinion of him. Tweak, ain't you ashamed of yourself?"

The canine hung his head as if conscious he had

done something not very creditable, and when Eustace added some sharp words of reproof he turned about and slunk away with his tail and head drooping.

"Well, Eustace, this is such a charming day, and there is so much of the afternoon before us, I don't suppose you want to go into camp?"

"No; that would be only a waste of time. We will push a little further to the mountain ridge yonder, over which we must pass, and to-morrow we will be on the southern water-shed of Java; then we shall not have very far to go before reaching the Sacred Mountain, and what are the most wonderful ruins in the whole world."

"What are they called?" asked the astonished Hermon.

"The great temple of Bara-Budur is the most marvelous. It is in the province of Kadu; but we expect to see more than that."

"What is there so amazing about Bara-Budur?"

"Its size, and the labor that must have been spent on it. It stands on a small hill, and has a central dome and seven ranges of terraced walls covering the hill, and forming open galleries each below the other, and communicating by steps and gate-ways. The dome in the center is fifty feet in diameter, and around it is a triple circle of seventy-two towers. The whole building is six hundred and twenty feet square and more than a hundred feet high."

"Are there any ornaments about the structure?"

"In the terrace walls are niches containing cross-legged figures larger than life."

"How many?"

"About four hundred, and both sides of all the terrace walls are covered with bas-reliefs, overflowing with figures, and carved in hard stone; you can see, therefore, that they occupy an extent of about three miles in length. Wallace, the traveler, says that the amount of human labor and skill expended on the Pyramids of Egypt sinks into insignificance when compared with that required to complete the sculptured hill-temple in Java."

While this conversation was passing between the cousins, they were walking leisurely in a southern direction. As there was nothing of special interest about the lake, they turned their backs on it, and it was soon left out of sight. The ascent was gradual, but the ground at times was so broken by bowlders and rocks that numerous detours were necessary, and the progress was not only slow but tiresome.

The vegetable growth was of tropical luxuriance, as it must of necessity be where it grows up to the very rim of the craters of the various volcanoes. More than once they observed the tall, symmetrical forms of the upas-trees; and Hermon, recalling the many strange stories he had read and heard of them, and his own experience, often stopped and surveyed them with great interest.

Except for the masses of stone, they would have little difficulty in traveling through the Javan forest. The trees, although numerous, did not attain a very large size. In some places the vines and shrubbery were dense, but not enough to interfere with any one passing between the trees.

Open spaces, where the graceful silvery grass was growing, were not infrequent, while the fruits and birds were a continual source of delight.

Parrots and paroquets, some of them as beautiful and almost as varied in their colors as the rainbow, flitted among the upper branches, and occasionally sat as motionless and solemn as owls, and looked down on the two passing beneath. When they indulged in their screeches they were as discordant as the noise made in filing a saw. The Java sparrows are as great nuisances in that country as their English cousins have become in the United States. They are always quarreling, and in some parts of the country drive away sleep for hours before daylight. They seem to be eternally engaged in wrangling or eating.

Other birds of variegated plumage were almost continually in sight, and Hermon was struck by the sight of several pea-fowl which ducked their heads and furtively watched the lads, who started them running along the ground or sailing away from the branches on which they were perched. This was done by the boys shouting at them and swinging their hats above their heads.

The peacock is certainly one of the most wonderful birds in the way of plumage (not even excepting the Bird of Paradise) that is found on the globe. It seems as if nature had set out to show what marvels she can perform in the way of decorating one of her creations.

"There," said Hermon, pausing a moment to pluck from a tree a fruit the size of a small orange, " is the most delicious thing in the whole world. I didn't particularly fancy it at first, but I am ready now to admit that it is all the Javanese claim for it."

The fruit to which he alluded is known in the East Indies as the mangosteen. It has a thick rind, something like the outside shell of a walnut, but within is a pulpy kernel which is of the most delicious flavor, and fairly melts away in one's mouth.

"Nearly everybody says as you do," replied Eustace, who also picked one and began removing the rind; "but when I was in St. Louis, Philadelphia and New York, I ate some apples and pears which to my taste are better than the mangosteen. If there is anything which grows that can beat a dead-ripe bell-flower apple or a juicy Bartlett pear, why I have never tasted it."

"I don't think there can be much improvement on them. Haven't you any apples in Java?"

"No apples grow in the East Indies. Further up in the mountains we may come across peaches and Chinese pears, while you have seen plenty of the tamarind-trees. Of course in the tropical countries we expect a greater variety of fruits, but there are several which grow in America that I wouldn't give for all of them. *Gracious!*"

Eustace was several steps in advance and to one side of his cousin when he uttered this exclamation and bounded backward. The startled Hermon saw what caused his fright. A serpent, nearly four feet in length, lay coiled directly in front of the other, evidently waiting till he came within striking distance. The serpent was slim and mottled on its belly, with red and greenish patches on its back. Faint-colored rings were around its tiny, glistening eyes, so that in that respect it suggested the "spectacled" cobra of India. Its crimson tongue was darting in and out, and its small head, elevated about a foot from the coil, and thrown back so that it resembled the neck of a swan, would have shot forward like a flash of light the instant its victim was close enough to make the blow a sure one.

There could be no doubt of its venomous nature. It was one of the most dangerous of the score of deadly serpents found in Java.

Without either of the boys speaking, Eustace

brought his rifle to his shoulder and took careful aim at the head of the reptile, which was not twenty feet distant. The hideous poison-fountain remained motionless, while the beady eyes gleamed like points of flame.

With the sharp crack of the rifle the head of the snake vanished as though it had never been. Eustace had shot it off and left the body, which began whipping the earth and turning over and over with bewildering rapidity, darting hither and thither, until it shortly settled to rest.

CHAPTER IX.

THE CAMP-FIRE.

"THAT'S the worst of these tropical countries," said Hermon, when they were moving forward again. "With all their beautiful flowers, fruits, birds and vegetation, they abound with serpents. Ugh!" he muttered with a shudder; "if there's one object in the world that I hate above all others it is a snake."

"But I suppose they serve their purposes like everything else."

"No doubt; and, according to my thinking, the best use they can be put to is to kill them. If I could have my way I never would allow a single one to be shown in the museums or anywhere else. I believe nearly everybody is born with a loathing of all kinds of snakes."

Without entering upon any argument (for, to tell the truth, Eustace shared the prejudice of his cousin, as you and I do), the elder whistled for their dog Tweak. There really was little reason to do so, for at that instant he came running at full speed among the trees toward them, giving expressions to an occasional bark of terror, and looking very much in earnest in his flight.

"There's some animal chasing Tweak," said the astonished Hermon, stopping short and gazing in that direction.

"Not one, but a number of them. They're wild hogs!"

A half-dozen swine, dark-colored, slim, bony, and fleet of foot, were galloping hard after the terrified hound, and emitting grunts of anger as they clung pretty close to his heels. In nosing through the forest he had roused some of them, who set upon him with such vigor that he was put to instant flight.

Now I am quite sure you know that the wild boar is one of the most formidable animals known, and, like the royal Bengal tiger, he is afraid of no living creature. He can run like a deer, and his amazing strength enables him easily to rip up any animal he can reach.

The wild pigs which were chasing Tweak were far from being the equal of the famous wild boar of Germany, but they possessed some of his characteristics, one of which is their undaunted courage.

The dog, finding his pursuers did not leave him, ran straight to Eustace for protection. Darting behind his master, he crouched down and whined with fear. This, of necessity, brought the young hunters to the fore.

The leading hog stopped short, looked up at the lad, who had raised his gun to his shoulder, and with an angry grunt he lowered his head again like a bull about to charge and made for him. His tusks were several inches long, and he could have rent the youth with a single blow had the opportunity been given.

But Eustace fired at once, and stretched the brave hog on the ground before he took three steps. The second member of the *Sus* family was laid low an instant later by Hermon, and the boys opened on them with their revolvers. There were but three of the wild hogs left, and they were demoralized by the vigor of the assault. They stood with their snouts thrown up, grunting and looking about, as if suspicious that there was a mistake somewhere. In a moment one of them squealed piteously, and dropping on his side and kicking the air with great vigor, departed this life. Thereupon the others broke and scampered off in the woods at the top of their speed.

The moment they started Tweak dashed after them, barking furiously, as if challenging them to stand their ground and fight. He was heard for several minutes after all disappeared, but soon came back, treading the ground with the air of a conqueror.

"That dog is the greatest fraud in Java!" exclaimed Hermon, disgusted with the cowardice he had shown from the beginning. "I can't understand why you brought him along or why you consent to own him!"

"I must say he hasn't shown up very well so far" said Eustace with a laugh. "But wait until we are through this jaunt, and you may change your opinion."

Hermon shook his head.

"He's too far gone to reform. I consider him a miserable cur, hardly worth ammunition to shoot him; but he is yours, and I'll let him alone out of regard for his friend."

Eustace laughed and begged his cousin to suspend his opinion until later on.

One thing caused growing uneasiness. Several times during the afternoon they caught sight of monkeys swinging among the limbs, chattering and showing much interest in them. Although there is little if any difference in the appearance of these animals, the boys were able to identify a number, and there could be no doubt that the leaders, as they may be called, were following the boys.

They were first noticed when the cousins left the volcanic lake, as it may be called, and had been scarcely out of sight since. The largest and oldest monkey was curiously marked about the shoulders with some patches of gray, so there could be no mistaking *him*.

When night came, and the lads kindled a fire in

the depths of the forest, there were fully fifty monkeys gathered among the tree-branches around them. They were swinging back and forth, running up and down and among the limbs, and indulging in all the comical antics natural to the creatures. Hermon stood with his back to the camp-fire, watching them as well as he could by the reflection of the blaze behind him. Frequently his merry laugh rang through the forest arches; for what boy can study the whimsical performances of a lot of monkeys without giving away to mirth?

But his cousin was thoughtful, evidently sharing the feelings of Tweak, who crouched at his feet, his instinct teaching him there was danger in the air.

"I tell you, Hermon," said he, as he stood close by his side, "we're in more peril than we have been since we left home."

"What do you mean?" asked his astonished cousin.

"Those monkeys are getting ready to attack us. That gray-back and the largest and oldest of the rest of the crowd have been following us ever since we left the lake. They have called in others, till you see what a gang they have got together. Such a force of monkeys have driven out a strong force of hunters more than once, and it isn't likely they will have much trouble with *us*."

"Why do they seek to disturb us, when we haven't molested them?"

"They seem to get the idea that we are invading their homes, and self-defense requires them to slay us."

"Well, if they do, we shall have to open on them with rifle and revolver, as we did with the wild hogs."

Eustace shook his head.

"You can't scare them off in that way. Look there!"

"The gray-backed monkey, as he was called, had swung down from his perch, and with three others as large and formidable as himself was cautiously advancing toward the boys. He was grimacing and chattering, and the odd twitching of his thin, black lips showed his yellow teeth, while his black eyes flashed with wrath.

"They mean to attack us," said Eustace. "Hold your gun ready; I'll drop the second one, and you shoot the gray-back that's in front."

CHAPTER X.

A PERILOUS SITUATION.

THE SITUATION of Eustace and Hermon was critical indeed. The monkeys were on every side, and beyond question were making ready to assail them. It looked as if the major part were timid about doing so, and the large gray-back and his three equally formidable companions were taking the lead.

The activity and superior intelligence of the creatures made them the most dangerous of foes. The leaders could not fail to understand their advantage, and knew how to use it.

There seemed but the one thing for the boys to do. If the monkeys assaulted them, they would defend themselves as best they could. There was some ground to hope that if the leaders were driven off the rest would not molest the boys.

"Don't fire till you have to," said Eustace, narrowly watching the mischievous animals. "Those four are searching for a chance to take us unawares."

"I'll let firelight through him!" exclaimed Hermon, holding his rifle ready. "He's the captain of the gang, and deserves hanging. Let me attend to his case."

Tweak had been terrified more than once since this jaunt began, but he now lost his senses entirely. He crouched whining at the feet of his master, until his terror became so excessive, on seeing the creatures steadily advancing, that he sprang directly at the gray-back, uttering a sharp bark. The instant he landed in front of the creature he attempted to leap backward, but was too late. The oddly-marked monkey with wonderful dexterity caught him in his arms as if he were a baby, and all three scampered off in the darkness.

"Good!" laughed Hermon. "They knew all the time what a valuable canine Tweak is, and they have been awaiting a chance to steal him. Now that they've got him, they're satisfied."

"And I shall be more than satisfied if such is the case," said Eustace, hoping rather than believing that his cousin might be right in his odd theory. "It's a sad affliction, of course, to lose so valuable a prize, but it is buying our lives with a small price if such proves to be the case."

"Yes; I don't see how the price could be any less."

But it was no time for jesting. They listened and looked in every direction.

When the quartet of larger and braver monkeys were stealthily advancing from the woods the others were emboldened by the sight, and began also creeping toward the young hunters, standing at bay by the camp-fire. Most of them, however, were among the branches, ready to pounce down on the boys whenever their courage was sufficient.

But when the four darted back in the gloom, one of them carrying the dog, the others did the same, so that for a brief time thereafter nothing was seen of the animals, though their chattering and peculiar cries were louder than ever.

"I wonder whether they will be satisfied with Tweak?" was the inquiring remark of Hermon, who all the time was watching all sides. "What will they do with him?"

"How can I tell?" asked Eustace, by way of reply. "No doubt they will put him to death; but monkeys are such peculiar creatures they are likely to have considerable fun before they dispose of him."

"Don't you think that the capture of Tweak will give them more courage, and lead them to attack us?"

"I shouldn't wonder, though it looked to me as if they would have done so in a very short time without any regard for him."

"But those four were the only ones that dared lead; but when the attack once opens the others will join in and what can we do against so many of them?"

"Can't we make some use of the fire?" asked Hermon; "you know that all animals are afraid of fire!"

"I have been thinking of that; but if we keep them off until morning they will have no fear of it then."

"If I had known these pests had such an objection to us I would have hunted for some route ——"

Hermon was cut short in his speech by something violently falling upon him from above. At first he thought the heavy branch of a tree had struck him, but it was the hound Tweak, which came down with a thump that knocked the breath for a moment from the youth's body and narrowly missed throwing him into the fire.

In fact it hurt Hermon much more than it did the dog. He rolled off the lad's shoulders, and alighting on his feet, uttered a whine of fear and crouched again by his young master. A glance at the canine showed that he had not been injured during his brief captivity among the monkeys. One of them had carefully climbed out among the limbs, until directly over Hermon, when he dropped the dog on him.

Instantly the woods echoed with the cries of delight on the part of the creatures, and Eustace would have broken into uproarious laughter but for the fact that he saw the serious side of the situation. He caught sight of one of the leaders in a squatting position, a dozen feet behind Hermon, and making ready to leap upon him. At the moment Eustace raised his gun the monkey started with incredible swiftness toward the lad, who, having partly turned, failed to see him, but was looking at something directly to the rear of his cousin.

Before the treacherous monkey could reach his enemy Eustace sent a bullet clean through his skull, and he fell as dead as a "door-nail."

"Look out, Hermon——"

But the elder observed that his companion had his rifle leveled at something behind his position, and he glanced furtively over his shoulder at the instant the sharp crack sounded through the woods.

Eustace was in time to catch sight of the huge gray-back that had attempted to play precisely the same trick on him when he was effectively checked by the bullet from the well-aimed gun of Hermon. Fortunately the boys did the very best thing possible. The sound of Hermon's gun was yet in the air when, dropping their rifles, the two whipped out their revolvers and began blazing away at the nearest animals.

Some of these were so close that they were struck; but, as was to be expected, the majority of shots went wide of the mark. The pistols were quickly emptied, and then the guns were snatched up again. You know how speedily a breech-loader can be reloaded, and in a twinkling, as it seemed, the sharper crack of the guns resounded among the

The monkey dropped the dog on Hermon Hadley's head.
(See page 97.)

trees. When they were charged a second time, and the youths glanced about for targets, not one was to be seen.

It is impossible to picture the panic created by the vigorous resistance of the boys. It cannot be doubted they were in the most imminent peril, for a combined and intelligent assault by the horde of monkeys could not have been successfully resisted. The young hunters would have killed a number of their assailants, but when their weapons were emptied of their charges they would have been defenceless.

As I have shown, the huge, spotted-back monkey and three others were the directors in the campaign. They were at the head, and they led the charge with the expectation that the rest would promptly follow up the attack, as they would surely have done; but the more timid animals, while looking for a success on the part of their officers, as they may be called, saw them shot to death with a quickness that took away the breath of the spectators.

While they were staring stupefied, the fatal bullets began whistling among them, and several monkeys, with frantic squeals, doubled up like jack-knives and took a header from the limbs on which they were perched. The others scrambled off in the darkness, terrified to distraction. Thus it was that when the boys with loaded guns in hand glanced around, they failed to catch a glimpse of a single monkey. They had not only retreated, but had become absolutely quiet. The stillness which reigned through the woods was profound.

"I do believe they are all gone," said Hermon, after he had turned completely around, with his eyes searching the limbs and the ground so far as the glow from the fire would permit. Eustace had gone through the same movement, each ready to discharge his rifle the instant a target was seen.

"That is true; but won't they come back again?"

"You can answer that question better than I, for you have seen more of the *simiadæ* family, and know more of their habits."

"As I told you a few minutes ago, there's nothing which one or two men in the woods dread more than a troop of combative monkeys. They are sure to be mischievous, and will play all sorts of tricks. Sometimes they become angry without any cause, and pester the hunters until they are compelled to retreat to save their lives. If the men kill one or two of the creatures the others become enraged, and an overwhelming attack is almost certain to follow.

"Why wasn't it so in *our* case?"

"We did up the thing too thoroughly. If we had slain but a single monkey, the rest would have been at us in a twinkling; but we opened on them in a style which has thoroughly scared all the others."

"But will they stay so?"

"I think we are safe, if we are vigilant. They

will not disturb us for a considerable time to come but a monkey is revengeful, and after a time some of the bolder will prowl around to see whether there isn't a chance to do us injury. It will take them a considerable while to get up enough courage, but if we should go to sleep they wouldn't hesitate long in attacking us."

"Go to sleep?" repeated Hermon. "I don't feel as though I could sleep a wink for a week."

"Because you feel so *now* is no reason to believe you will feel so an hour or two later."

Hermon shook his head.

"I have had too good a shaking up to sleep any between now and daylight. I never knew how heavy Tweak was until he dropped on my head and shoulders. It seemed to me, when he struck, that he weighed a ton! My shoulders still ache. What an idea!" exclaimed Hermon with a laugh, as he rubbed one shoulder and looked down at the dog, which was stretched out with his nose between his paws, as if all danger was over.

"The monkeys are original in their ideas. I once heard a clergyman say that he has stood by a cage of them for hours and enjoyed himself as he could nowhere else in watching their comical tricks. I never knew of one of the creatures dropping a dog on the head of a person, but I am quite sure it was done many a time, if such a good chance as this was given."

"We must keep the fire going," remarked Hermon; "and we can't do that without fuel."

When the cousins went into camp for the night they gathered enough wood to last for an hour or two. Where wood was all around them it was not to be supposed there could be any difficulty in collecting all they wanted; nor, indeed, was there any such trouble, but the boys felt much misgiving on account of the monkeys.

"We won't have to go far to get the wood we want," said Hermon, "and I'll do it if you guard me with your gun."

As this was evidently the wisest course, it was adopted. The fire was stirred so that the circle of illumination was greatly widened. As the gloomy depths were lit up the boys caught a glimpse of several monkeys among the limbs, but they skurried out of sight. This proved what had been suspected. The resentful creatures were on the watch, and were sure to seize the first occasion that presented itself.

Hermon left his gun behind, as it was likely to be an incumbrance while gathering wood, and would be more useful at the command of Eustace, who was a quick and good shot.

The young American could not find much dead and decayed wood, which was necessary when he started the camp-fire, and really there was no need of it. The coals were so alive and glowing that they took hold of and consumed that which was green.

CHAPTER XI.

GATHERING FUEL UNDER DIFFICULTIES.

IT TOOK genuine courage for the young American to venture out in the gloom after wood when he was certain that more than one fierce animal was eagerly awaiting a chance to pounce upon him. Hermon did not mean to go beyond the circle of illumination, and he knew Eustace was covering him as best he could.

While cutting and breaking off limbs, he could hear the rustle, now and then, as the monkeys moved stealthily to and fro. Several times, too, he caught a glimpse of the animals, and instinctively drew back and placed his hand on his revolver.

"Have a care! don't go too far!" called Eustace from the camp-fire.

Hermon had brought in four armfuls, and intended that the next should be the last. He had gleaned the field around him so closely that he stepped a little further away than was prudent, and his cousin, who was closely watching him, warned him to be careful.

Hermon shouted back that he was going no

further, and began breaking off some limbs which hung almost to his shoulders. He had thrown several large pieces on the ground when a peculiar, snarling growl was heard. He looked searchingly around, but could see nothing.

"Eustace, stir up the fire a little," he called, dropping the branches from his hands and drawing his pistol; "I can't see well enough."

His cousin did as requested, and the yellow light pierced the gloom. By its glow Hermon perceived a large monkey squatting on the ground less than a dozen yards from him. He was directly in front, and seemed swollen to the exploding point with rage. He was as large as one of those that had acted as leaders of the troop, and would prove an ugly customer in a fight.

Had he kept quiet, he might have leaped upon the boy, and caught him at a great disadvantage; but he was too wrathful to contain himself.

As dimly seen, he was a hideous object, and the youth shuddered. He was so nigh, indeed, that Hermon was sure he could kill him with a single shot from his revolver; but he hesitated. There was something in the act which grated on his feelings, and he decided that if he left him alone he would not disturb him.

But the situation of Hermon was a peculiar one. It was necessary for him to retreat without his fuel or else to throw aside his weapon temporarily, for both arms were required to carry the bundle of wood to its place beside the camp-fire.

When the darkness was lit up by the added glow from the flames that Eustace stirred, the threatening monkey moved back a few steps; but he still faced the lad, and at brief intervals emitted that peculiar snarling growl which told his unsmotherable rage.

After confronting him for a minute or two, Hermon mocked his cries and took a step toward him. The monkey stopped his noise on the instant, and withdrew so far that he could not be seen.

That settled the question which Hermon had been asking himself. Angry as the monkey was, he had not the courage just then to attack him. He therefore shoved the pistol back in his hip-pocket, and, stooping over, hastily gathered up the wood he had broken off; but while doing so you may be sure he kept a close watch for his enemy whose savage growls were heard again.

The youth was determined to have his last package of wood because the monkey threatened him, but when he straightened up with his arms clasping the huge bundle he could not but reflect at what a disadvantage he was placed in case his enemy made a sudden assault upon him.

In going toward the camp-fire, Hermon did it after the manner of a subject withdrawing from the presence of royalty. He kept his face turned toward the monkey, and stepped backward with solemn deliberation.

He knew, or at least suspected, that the animal would become bolder than before, if, indeed, he did not make an attack; for nothing gives such courage to man or beast as the sight of a retreating enemy.

Sure enough, he was no more than fairly started when the monkey, with a louder growl than ever, came out of the darkness and ran swiftly toward him. He had resolved to assail the person that had inflicted such injury upon his kindred.

The sight of the frightful creature approaching with such celerity startled the young American, who hesitated for a second as to the best course to pursue. Had he dropped the wood in his arms on the instant and drawn his pistol he could have riddled his assailant; but it occurred to him that he ought to reach the camp-fire with the least possible delay.

Still facing the foe, therefore, he began stepping rapidly backward. At the second or third step his heel struck some running vine, which was like steel wire, and he went over on his back with his heels in the air. At the same instant the monkey, with a screech, made a bound for him.

Fortunately indeed was it for Hermon Hadley that when he fell so unexpectedly he did not let go of the bundle of sticks. Had he done so, he must have suffered grievous injury; but instinctively he threw up his arms, keeping them still clasped around the wood, which made the best possible shield for his face and the front part of his body, though the hands were exposed.

At the moment he fell the monkey landed on the bundle and began fiercely clawing to get at his face; but the usually cunning creature forgot that, a few steps away, by the side of the blazing camp-fire, another of the hated race was standing with rifle leveled at the savage assailant, who was in plain sight.

When the sharp report of the gun resounded through the woods it was the death-knell of the infuriated monkey.

CHAPTER XII.

STEALING A MARCH.

HERMON was not injured by the spiteful assault of the monkey. His aim was to reach his face, and when he alighted on the bundle of wood he began tearing it apart without paying any attention to the hands within his reach.

The plucky youth still girdled the mass of wood with his arms, and, though it was awkward and inconvenient to do so, he managed to struggle to his feet without dropping a stick. Then he hastily joined his cousin, who was in the act of reloading his gun.

"Eustace," he said, some minutes later, "I believe when morning comes the monkeys will make another attack on us."

"I feel certain they will; but we can give them a hot reception."

"It seems to me the number is increasing, and there may be so many at daylight we can't stand before them."

"That is likely to be the case, and we shall be fortunate if we can withdraw to the nearest settlement."

"Why can't we flank the pests?"

"How?"

"Let us steal away from camp in the darkness, and by daylight we may be beyond their reach."

"I never thought of that. I don't know whether it is possible or not. How can we prevent them seeing us go away?"

"I don't know as we can, but we won't lose anything by making the trial."

That boy may well congratulate himself who succeeds in outwitting a fox or a monkey; but Eustace and Hermon, after looking over the ground, as may be said, decided to make the attempt. Their plan was simple. They would try to steal from camp unobserved by their enemies. When fairly beyond the lines, they could hasten away for several miles, and place themselves entirely beyond reach of the animals.

It would seem there was little ground for hope; but, as Hermon observed, they had nothing to lose and everything to gain by the attempt.

"We can only succeed," said Eustace, with a half smile, "by leaving Tweak behind."

"That is one strong inducement to make the effort," replied Hermon, looking down with contempt on the dog, stretched out and sound asleep, with his nose still between his paws.

"You are uncharitable toward Tweak," said the elder. "The game which he has been forced to face is more than a bloodhound or the biggest bull-dog could overcome."

"I would like to see a monkey carry off one of those dogs and drop him on the head of his master! If he tried it, it would be the monkey that would drop!"

"Perhaps they may make a pet of Tweak," laughed Eustace, "for there is no guessing the whims of such creatures."

It will be understood that the dog, unable to comprehend what the boys wished to do, would be sure, if taken along, to betray their presence by his actions. Therefore it was necessary that he should be left behind.

In order to succeed, the fire was allowed to smoulder until objects were invisible at a distance of two or three yards. They hardly would have dared to do this had not the animals remained so quiet that the lads were almost certain they would not attack them before daylight.

Having agreed upon their course, Hermon, who lay on the edge of what may be called the line of visibility, noiselessly moved away, until any one viewing the spot would fail to note the least sign of him. An American Indian could not have withdrawn from an enemy's camp with greater skill than he.

Not the slightest sound or signal was uttered by either boy, for it was not required. Eustace listened for that which he did not wish to hear—the sounds indicating the discovery of his cousin's movement on the part of the monkeys. They would be sure to give evidence of such discovery the moment it was made, and Hermon would be compelled to return and stay by the fire through the rest of the night.

Eustace Hadley was surrounded by impenetrable darkness when, believing he was at a safe distance from the smouldering camp-fire, he paused in his stealthy movement and listened for some signal from Hermon, who had preceded him by half an hour.

The faint, chirping sounds which betrayed the proximity of the monkeys were heard no more, and he was certain he was far beyond the entire party which had threatened within so short a time the lives of himself and cousin.

As it was impossible in the darkness to see each other, the friends agreed before leaving the camp that so far as it was in their power they would follow a certain fixed direction. The task of Eustace was the more difficult because the fire was nearly gone when he left it, and he could only rely upon conjecture to determine the right course to follow.

Both boys understood this from the first, and Hermon had been asked to signal to Eustace when the proper time came to do so. That signal was a faint, tremulous whistle—so faint, indeed, that no one beside Eustace would suspect it was not made by some bird.

Failing to hear it, Eustace softly whistled and awaited the reply. Not, however, until he had uttered the signal a second time did he receive the welcome response. His heart started with delight.

"It begins to look as though I had no cause for fear," he thought, stealing toward the point whence issued the call.

The fact that the signal came from a point on his left was another proof of the impossibility of any one following a straight line without the aid of his eyes or artificial help. He had started directly behind Hermon and believed he was advancing in the same course, whereas their paths widely diverged.

Only a rod or two were passed when Hermon called again, and Eustace found they were so close that he ventured to speak in a guarded undertone.

"Halloo! is that you, Hermon?"

"I think so; move a little to the left."

"That's what I call a piece of the best kind of luck," added the elder, advancing at rather a reckless rate toward his friend; "it isn't often you can flank a lot of monkeys—*confound it!*"

The outreaching hands, one of which held his gun, somehow or other missed a limb which glided over them and directly under the chin of Eustace, who was almost lifted off his feet. He speedily disentangled himself, however, and, though the sensation was anything but pleasant, joined his friend in a laugh, which was made the more hearty because of their remarkable escape.

"You haven't got that dog with you?" asked Hermon, when they had congratulated each other in the impenetrable gloom.

"No—sad to say, I was forced to leave him behind; but don't take it too hard, Hermon; sorrow must come to us all sooner or later; try to forget it."

"I'll do my best," replied the younger, with mock seriousness, "but it's mighty hard—that is for the monkeys, if they think they have got a prize; but, Eustace, here we are, and I don't suppose it's your idea to remain till morning?"

"No; we're too close to the monkeys."

"What do you think they will do at daylight, when they get ready to scoop us in and find we have gone?"

"There will be a great howling and outcry; they will mourn over their fallen ones, and lots of them will scamper back and forth on the hunt for us."

"Can't they follow our trail?"

"No; a monkey isn't worth a cent at that—that is, I have never heard of them doing anything of the kind; they will search through the woods perhaps for several hundred yards, and then, failing to find us, will give it up."

"But it is best we should put a good distance between our last camp and the point where daylight finds us."

"There can be only one opinion as to *that*."

"You know more of these woods than I do and will have to take the lead."

"I don't know enough about them to make any better guide than yourself," replied Eustace; "a native that had spent years in hunting through this section couldn't help us where it is so dark."

"Halloo! there's the moon!" exclaimed Hermon, at the moment an opening among the branches overhead gave sight of the orb, as it came out from behind some clouds.

"That will help us," said his friend; "but there is so much vegetation that the shadow is almost everywhere, and we shall have to be careful where we walk."

It was very tedious indeed, but neither lad complained, for the necessity of such flight was too evident for either to lag. Now and then they reached open places, where the ground was covered with the long grass of which so much had been seen during the day. Across these they walked with more freedom and swiftness.

The stillness which reigned most of the time was broken now and then by the call of some night-bird, and occasionally the cry of a wild animal floated on the night-air. Three different times a strange, piercing sound reached their ears, which Eustace declared was made by a tiger—possibly the mate of the one that had been killed the night before.

While crossing an opening more than an acre in extent, which was broken by bowlders and rocks, a number of immense birds darted back and forth over their heads, and in irregular lines.

"They act just like bats," said Hermon, who struck at one with his gun, missing it by a couple of feet.

"They *are* bats," replied his cousin; "and when we show bats in Java, they are worth looking at."

"They beat anything I ever saw; how big are they?"

"There are plenty of them which measure five feet from tip to tip of their wings; can you beat that in America?"

"No; we never allow that kind of animal to have wings. When anything does, we call it the eagle; and the American eagle, as you know, can beat all creation."

"The popular name of these creatures," said Eustace, "is the flying fox."

"Why are they given such a name?"

"Because of the red, fox-like color of the fur and the shape of the head, which is almost exactly that of a fox."

"They are big enough to be pretty respectable enemies."

"They don't amount to anything that way, but they often make havoc among the fruits of Java, and have almost ruined the crops in some parts of the island. They are the oddest of creatures; they will hang by the score from the trunks or limbs of trees, and are so quiet that if you came upon them alone you would take them for some kind of fruit. If you disturb them they will scream and flutter helplessly about, for their eyes are good for nothing except in the night-time."

"It is a wonder we did not pass some of them during the day."

"I have no doubt we passed hundreds, but you might go within a dozen steps without noticing them. If you could see the head of a flying fox as he squats on the ground, resting on his hind feet and elbows, you would be as much surprised when he rose in air as you would to see a lion go skimming toward the clouds."

By this time the boys were confident they were a full mile from the place of their last camp. Well aware as they were of the certainty of going wrong, and the probability that they would travel in a large circle and ultimately come back to the point from which they started, they guarded very carefully against such a calamity by frequently consulting their pocket-compasses. At intervals these were taken out and examined by means of a lighted match. It caused many expressions of wonderment, for every time they did so they discovered they were swerving from the right course. Occasionally this was so marked that the first one who looked at his compass was certain it had become disarranged; but when the two were found to agree the question was settled, and the boys were not foolish enough to disregard what the instruments said.

At two o'clock in the morning the lads agreed that fully a couple of miles had been placed behind them, and it was safe to hunt some spot where they could rest until morning.

The place was so inviting that it was decided to use it for that purpose. It was on the border of another of those natural clearings of which mention had been made several times. An enormous mass of volcanic deposit, which must have weighed many tons, was heaped together, and along the side ran a small stream of clear, pure water.

In venturing on their jaunt through Java the cousins had made no such preparations as they would have done on entering upon a hunt in America. In the first place, there was nothing in the way of severe weather to be feared, for it was never known at that season of the year. Being the most thickly-populated country in the world, they expected always to be within hail, as may be said, of some settlement; and with their compasses and Eustace's general knowledge of the island they really could not go far wrong.

True, there are heights to which the central mountain range rises which reach an arctic temperature, and where, even though it lies so near the equator, one can readily freeze to death; but the young hunters had no thought of climbing such summits, and they meant to pass over the dividing ridge between the rising and the setting of the sun.

But when they agreed to stop where they were they leaned their guns against the rocks and began collecting wood with which to start the fire. While there was no particular need of this, it was a great comfort to have it. There is something in the cheery, roaring blaze, in the depths of the wilderness, which does a great deal toward dispelling the gloom of the solitude. It is a pleasure to look in each other's faces when speaking, and the spot becomes ten times more cheerful by reason of its sharp contrast with the surrounding darkness.

It might be that the glare of the flames would summon some of the wild animals from the woods. Perhaps the monkeys or lutungs (black apes) would be drawn around them; but the fire was started in such a spot that it was almost entirely sheltered by the rocks, and could not be discovered from more than one or two points in the forest.

Accordingly, the sticks were heaped up against the solid wall and the match applied. In a few minutes the flames flashed out, and the boys sat close together and congratulated each other on the good fortune that had attended their efforts to "flank" the troublesome monkeys.

"I didn't believe either of us would succeed," said Eustace, "for we know how cunning those animals are. I was sure every minute you would come rushing back to the camp-fire with two or three of them on your shoulders clawing at your head; but when my watch showed me a half hour had gone by I knew you were safe."

"Then you had little doubt you would succeed yourself?"

"It would seem I ought to have had no doubt; but when I crept away from the fire and rose to my feet in the darkness beyond, something told me I was going to fail."

"Something told you," repeated Hermon incredulously. "What's the use of talking that way? What could have told you? I have heard good old ladies discourse in that style, but you and I know it's all superstition."

"It may have been, but the feeling was there, all the same. What do you suppose was the cause?"

"Your heart was probably oppressed with grief because you had to leave Tweak behind. I only wonder that you didn't take him in your arms, just like that other monkey, and run the risk of losing your own life rather than part company with the fraud!"

Eustace laughed.

"I think you have had enough to say about that poor cur to let him drop——"

"He has had one drop already, and that's enough for me," interrupted Hermon, rubbing his shoulders over the remembrance.

"He is no longer a factor in this business," said Eustace, with mock seriousness. "He is eliminated from the problem, so to speak, and I therefore forbid all mention of his name hereafter, unless it be in the respectful manner that should be used in referring to the departed."

"I will endeavor to respect that decision," remarked Hermon, with a solemn inclination of his head.

CHAPTER XIII.

AN ALARM.

EUSTACE and Hermon were quite tired from their laborious tramp through the woods, and when they lolled on the ground in easy postures the rest was so grateful that they soon became drowsy and laid their heads down in slumber.

It cannot be said this was a wise proceeding, for despite the favoring features of the Javan forests, which I have shown, the recent experience of the young hunters proved that an element of danger was always present, and the two should never have allowed themselves, when alone, to be unconscious at the same time.

Such had been the understanding between them; but what boy does not know how insidious an enemy is sleep? Haven't you and I many a time determined to keep awake through the night, or for a certain number of hours, and haven't we dropped off into the "land of Nod" just as helplessly as an infant? How often have we resolved we would take notice of the moment we fell asleep, and yet, when morning came, it was impossible to recall the time? Of course it always will be impossible, for the very best of reasons.

The boys were some hundreds of feet higher than when they left their former camp, and though within a tropical climate there was a perceptible coolness in the air which made their blankets acceptable. They lay with their backs together and their feet toward the smouldering camp-fire, and slept the sleep which comes to the healthy of body and mind.

But that kind Heaven to which both always appealed before lying down and on rising in the morning kept better watch over the unconscious forms than any human sentinel could have done. There was peril on every hand but it came not nigh them, and when they opened their eyes almost at the same moment the sun was above the horizon, the fire had long died out, and no animal or reptile was in sight.

They were somewhat stiffened from lying motionless so long on the ground, which in Java is generally moist, but their youth and vigor enabled them quickly to rid themselves of it. They carefully bathed their hands and faces in the clear, running water, and took one or two swallows of the refreshing fluid. Then they were ready to hunt for something to eat.

"How about those monkeys?" asked Hermon, as though he had forgotten them until that moment.

"No doubt they are ready to fall upon and rend each other in their disappointment over our flight.

If monkeys can talk—and I believe all animals have a language of their own—they must be indulging in very hard words."

"I presume the principal expletive they use is 'Dog gone it!'"

"No; because the dog isn't gone, or at least wasn't when they found we had taken our departure."

"But our respected and esteemed canine has wished they could say it—well, I'll be hanged!"

The astounded Hermon almost dropped his rifle in his amazement; for there on the edge of the woods were seen two grinning monkeys, hopping and skipping under the trees and running up the limbs.

Eustace was astonished too, and both were frightened, as they well might be; for if their enemies had succeeded in tracing them thus far, trouble was sure to follow.

Hermon was so exasperated that he leveled his gun at one of the monkeys, but Eustace pushed the weapon aside.

"Why do you want to do that?"

"If we have got to fight, we might as well open the ball. It was our promptness that won last night."

"But we are not sure they will attack us; we shot their leaders and the others may be afraid."

The boys, however, were so sure of being assailed by the enraged creatures that they stood a full half hour with the hammers of their guns drawn back, and ready to open battle.

But it was noticeable that the two monkeys first seen were the only ones that appeared. The boys supposed they were the skirmishers, as may be said, of the main body, which would soon swarm from among the trees; but as the time passed and nothing of the kind appeared, the truth gradually dawned upon the elder.

"Well, there!" he exclaimed, dropping the butt of his rifle on the ground with a bang; "that is a joke on us."

Hermon looked inquiringly at him.

"I don't understand you."

"There are only two monkeys, and they don't belong to the gang that molested us last night."

"Is it possible?" and Hermon let down the hammer of his gun and lowered the weapon.

"There can be no doubt of it. I propose now that we give our whole minds to the task of hunting for breakfast."

Despite the assurance of Eustace, his cousin was not wholly satisfied until they had tramped along the edge of the clearing and gone some distance among the trees. The monkeys followed them only part way, when, as if they considered them of no account, they vanished and were seen no more.

Had our friends been hunting in the depths of an American wilderness they would have craved a more substantial diet than mere fruit, for you know meat is heating to the blood, and in the extremely cold regions of the far north nothing else is used. If any of my young friends should spend a few months in the desolate lands visited by Lieutenant Greely, or even further south, they would learn to drink oil as if it were soda-water, and a tallow-candle would be as toothsome as a stick of candy.

And so it was that though Hermon and Eustace were vigorous youths, capable of undergoing as much fatigue as any of their age, they wanted only a fruit breakfast. The meat which they had eaten at the house of Jurak the morning before was all they cared to have for several days to come.

That fruit was at command in the form of the delicious mangosteen, which, as I told you some time ago, is considered by many to be without an equal in the world. They plucked it from the trees around them and ate until their healthy appetites were fully satisfied.

The boys were now near the middle of the Island of Java, as regards both the length and breadth. They had yet to cross the mountain ridge which stretches from east to west, and which at different points is broken by volcanoes, some of which are active, while others are at rest. There had been times when it would have been a work of extreme difficulty for any one to make his way over this chain, just as it was hard and dangerous a few years ago for a company to force the passage of the Rocky Mountains.

But for a long period the Javan range has been traversed by well-beaten roads which take advantage of natural depressions, so that hundreds of persons cross both ways every week on foot, on their small, tough horses, or in the rude vehicles of the country.

Eustace was in favor of turning to the right along the side of the mountain until they struck one of those roads, and then following the regular route to the other side; but Hermon thought it would be more interesting and more in keeping with their characters of amateur sportsmen to push through the forest.

"I'm willing," said the elder, looking toward the enormous pile of wooded country which towered in front of them; "but I want you to understand that there will be more work than play in it."

"Well, I'm not afraid of work," said the plucky Hermon. "We're out for something of the kind and ought to be able to rough it."

"Enough said," remarked Eustace with a laugh, as they resumed their tramp among the trees, across the open spaces and around the rocky masses that were often encountered.

The clouds which had partly obscured the moon the night before were now gone and the sun shone from a clear sky. For several hours during the early part of the forenoon the air seemed to be in equipoise, but after a time a gentle breeze stirred the vegetation and the temperature became decidedly cool.

"We would need a fire and our blankets to-night," said Eustace, "if we camped on the highest part of those mountains."

"Where do you expect to camp?"

"I can't say positively, except that it won't be on the summit of that ridge. You can see a deep depression just off to the left," added the elder, pointing to what is known as a "pass" in the western part of our country. It was several miles in width, and, intersecting the mountain chain as it did, it looked as if it marked the ancient bed of some mighty river which had burst its way from the Javan Sea to the Indian Ocean.

Of course all the highways for many miles east and west converged through this natural depression in passing from the low, flat, alluvial shores on the north to the precipitous bluffs on the south; but between such roads were stretches of wild forests and broken country in which are found many wild animals and birds, and through which it is difficult to travel.

Almost any one would have avoided that part of the pass and taken to the routes where traveling was so easy; but the young American insisted that if they wanted nothing but a comfortable time they ought to stay at home altogether. They possessed one cheering thought, however: at no time would they be far from civilization, and should they need friends they were quite certain of finding them.

Thus it came about that the middle of that afternoon found them fairly within the broad natural highway that had been opened centuries before by some prodigious convulsion of nature, with every reason to believe that on the next day they would cross the dividing-line and begin descending toward the Sacred Mountain, where they were to see the wonderful ruins.

It had been their hope to arrive there that day; but I have shown what happened to them, and as no necessity for haste existed, they were sensible enough to take all the time needed.

Though the sky was quite clear, there was visible at all times a dark, muddy-looking vapor, resting almost motionless in the horizon or stealing part way toward the zenith. Eustace informed his cousin that this came from the volcanoes, of which there are about two-score in Java.

Toward night, when the air had become still, the boys heard very distinctly a resounding sound, as if made by a heavy bell. The tones remained a long time trembling and gradually dying in the air, as they do when a heavy bell is struck a single time.

"That sounds like a gong," said Hermon.

"And it *is* a gong," replied the other. "Don't you remember the watchmen that we saw stationed a mile apart, each with a gun and a little house in which to shelter himself?"

"Yes; we saw several of them."

"Well, when they have anything to say to each other they do it by means of those signals. Hark!"

The response of the answering gong came faintly through the mile of intervening space, while the third, which doubtless took it up, was inaudible to the listening youths. Thus the signal, which probably started well toward Batang or Samarang, on the Javan Sea, would go leaping a mile at a time, until it stopped on the shore of the Indian Ocean.

It was not yet dark when the boys, finding themselves pretty tired, concluded to do the sensible thing, and instead of sleeping in the woods apply to some native for lodging or enter some town where they would have no trouble in finding accommodations.

"I remember when I was a wee bit of a chap," said Hermon, with a laugh, "that father brought home the carcass of a deer which some hunter sold to him. The venison was cooked for dinner and was about as tough as the legs of your boots, and I must say there wasn't much more flavor to it; but because it *was* venison I thought it was extra-nice, and a diet on it would make a great hunter of me; so I chewed away until my jaws ached and I was miserable clean through. Father saw what a dunce I was making of myself, and took mercy on me before I starved to death and gave me some roast beef. Now, we are so close to good accommodations that we would be equally foolish if we refused to take them instead of sleeping outdoors."

"There wouldn't be much difference—here we are!"

They emerged into a highway at that moment. It was about twenty feet wide, without fence on either side, and was well worn by the wheels of carts, the hoofs of horses and oxen, and the feet of men, countless multitudes of whom had traversed it from one side of the island to the other.

No one was in sight at the moment, but they had taken only a few steps when an arrival came upon them with the suddenness of a thunderbolt, and with equally terrifying effect.

CHAPTER XIV.

JURAK.

IN THE middle of the road suddenly appeared a man, naked to the waist, barefooted, with his long hair streaming, eyes glaring, and running at the top of his speed. He held in his hand a *kris*, one of those fearful Javan knives, straight-bladed and fully a foot and a half long, with which he sawed and circled through the air back and forth as he rushed forward. At every few steps he leaped from the ground, uttering the most frightful yells, and frequently glancing over his shoulder at those who were pursuing him with their utmost speed.

"What does it all mean?" asked the bewildered Hermon.

"He is running amuck; move back or he will attack us."

"If he attacks *me*, I'll settle him," muttered the young American, raising the hammer of his rifle. "I don't stand any nonsense, and when that sort of an animal comes for a fellow it's time to shoot off a gun."

The boys stepped nimbly to one side of the road, and they would have run among the trees, had they not known the crazy man had seen them, and such action would encourage him to make an attack. They kept their faces toward him as he approached, bounding high in air and emitting ear-splitting screeches, and held their guns ready for use.

The native was quick to perceive the youths, and with another curdling yell he swerved to the right and made straight for Hermon, his rapidly-circling knife flashing in the sunlight as it played around his head and shoulders.

"All right, old fellow!" exclaimed the youth, bringing his gun to his shoulder; "if that's what you mean, I'm ready!"

Dreadful-looking as was the frantic native, the lad did not wish to fire until it became absolutely necessary in order to escape the murderous knife, held in the grip of the lunatic. The latter ran swiftly toward him, until within several steps. Then he leaped upward, and as he came down wheeled and continued along the road at the same headlong pace. Disordered as was his intellect, he had no wish to run against the muzzle of a loaded rifle, which would have been discharged had he advanced a step further.

"That was the closest call *you* ever had," remarked Hermon, lowering his piece and looking after the flying fugitive, at whose heels were half a dozen pursuers.

One of the latter was fully as fleet as the native,

and during the cavorting of the latter he caught up to him. Suddenly the fugitive was tripped and fell to the ground with great violence. The one who threw him leaped upon him in a twinkling; there was a furious struggle, during which the two were half hidden by the dust, and then the knife of the crazy man flew a dozen feet from his grasp; he was disarmed, and his conqueror choked him until he ceased to resist and was on the point of suffocation. Then, when he was so weak he could hardly stand, he was helped to his feet, and with a couple of Javanese tightly grasping each arm he began walking slowly back toward the village from which he started to run his muck.

As soon as he regained his breath he began to struggle to free himself; but the four who held him did not loosen their grasp, and he was compelled to cease from exhaustion.

"He has become crazy from using too much opium," explained Eustace. "It isn't often that one of our natives is taken that way, but it is quite common among the Malays."

"How does the disease show itself?"

"You have just seen; the man may be as quiet as you or I, when all at once he becomes frantic; he starts off, knife in hand, slashing all within his reach. Sometimes they slay a number of people before they can be checked, for they make such a sudden start that very few persons are prepared."

"What is the usual course with them?"

"They are generally shot down, and any one who does so is held guiltless by the law. Had you killed him, no one would have blamed you in the least."

"I am glad I didn't, for it would have been dreadful. How was it that no one harmed him?"

"He has many friends, and a few daring ones made special efforts to run him down before he could do much harm—*heigho!*"

The native who had tripped the lunatic was not one of the four who held him by the arms. He was walking immediately behind them and giving directions. The above sentences had scarcely passed between the boys when he came so close that he turned his head and looked at them. As he did so he and Eustace recognized each other, and with smiling faces strode forward with outstretched hands. Saluting each other, they spoke some words in the Javanese, and it was plain they were friends.

"Hermon," said his cousin, still holding the hand of the native, "this is my particular friend Jurak, at whose house we stayed night before last."

"I am real glad to become acquainted with you," said the finely-formed native, speaking with a deliberate accent which was almost as good as that of a native Englishman. He pressed the hand of Hermon warmly and smiled broadly. The last did not add any attractiveness to his naturally ugly countenance, for, like his wife, his teeth had been stained a black color, and many of them were far gone in decay; but the American felt a strong admiration for him because of the pluck and deftness he had shown in capturing the crazy native that was running the muck.

"I am delighted indeed, Jurak," said Hermon. "Eustace has had so many kind words to say about you that I never would have consented to go home without making your acquaintance."

"You speak kinder words than I deserve," responded Jurak, who, it was plain, was much pleased by the compliment.

As the natives having in charge the crazy fellow started back toward the village with their prisoner they met others coming out to learn what had become of him. Some of these looked curiously at Jurak and the two youths talking together at the roadside; but none was guilty of the least disrespect, and in a brief while the whole crowd were moving away, while Jurak and the boys walked far enough in the rear to talk freely without interruption.

Jurak proved the most entertaining of companions, and Hermon could well understand why he was held in such high esteem by Eustace. He was unusually intelligent, and the elder youth told his cousin that he wrote English as well as he spoke it. This was the more remarkable, for the educational facilities in Java are meager; I am sorry to say the Dutch are more anxious to make money out of the natives than they are to improve the mental and moral condition of the people.

Jurak said that on hearing that a large tiger was prowling through the neighborhood a couple of days before he had told Mr. Hadley, his employer, who gave him permission to stay away several days, or as long as was necessary. When it was learned that the terrible beast had been killed by a mere lad, who fired from the hut of Jurak, the wonderment was great, for such an occurrence was unknown until then.

Jurak and a party of hunters reached the home of the native within a couple of hours of the departure of the boys. Jurak remained that day and night at home, but as it was near the close of the week, he concluded not to go to work till the following Monday.

The Javans divide time into weeks, months and years. Their week is either of five days or seven. The former regulates the market, and is most generally adopted through the country. The week of seven days is used only as referring to the seasons, and to keep a more systematic account of the time than could be done by means of the little shorter week.

"Perhaps," said Eustace, as they walked slowly toward the village, "you would like to hear the Javanese names of the days in the week."

"I would, though I never can remember them."

"They are *Dati*, Sunday; *Soma*, Monday; *Anggara*, Tuesday; *Buddha*, Wednesday; *Raspati*, Thursday; *Sukra*, Friday; *Sanischarah*, or *Sumpah*, Saturday."

"How long is their year?"

"They use the lunar year; that is, three hundred and sixty-four days."

"I don't suppose they reckon time from the birth of our Saviour?"

"Of course not, but from the arrival in their country of a being whom they call Aji Saka. That is said to have taken place seventy-four years after the commencement of the Christian era; so that you must subtract seventy-four from our year to find the corresponding Javanese year."

"What is to be said of the religion of this country?"

As Eustace hesitated, Jurak took upon himself to reply to this question.

"There are many legends about the early history of Java, but I cannot think they are truthful; indeed they cannot be. The first white people who visited it were the Portuguese, who came here in the year 1510, as the Christians reckon time. The inhabitants of this and the neighboring islands were found to be well civilized. They were devoted to agricultural pursuits, built vessels, and carried on a large trade with the islands around them. The Dutch first visited Java in 1595, and they settled here a few years later. The natives at that time were Mohammedans."

"Have they always been such?"

"No; the records preserved by our nobles and the many monuments to be seen all over the island show that before the Mohammedans their religion was that of Buddha, which you know prevails in Hindostan."

"And before Buddha?"

Jurak shrugged his shoulders, turned the palms of his hands upward and shoved them away from his body, much as the Frenchman is supposed to do when puzzled, and shook his head.

"Who shall tell?"

By this time they reached the edge of the native village, which was attractive and picturesque. All the houses were made of split bamboo, and each was surrounded by luxuriant vegetation. The dwellings, as I have told you before, do not as a rule have any windows. There is no winter in Java and the people live almost entirely outdoors, their huts being mainly useful as storage-rooms for their few possessions and to cover them when they sleep. The ordinary dwelling-house costs from a dollar and a half to three dollars, so that the head of every family owns the house in which he is said to live.

"The Javanese are fond of society," explained Eustace, in answer to the inquiring looks which Hermon cast about him, "and they nearly always build in villages. Even though Jurak lives by himself, you see he has his friends here, and he visits these people a great deal."

"And so does Myeta," added the husband; "she will be here to-morrow evening, and will stay till Monday."

"You will see," continued Eustace, "that in forming a village the people take care to have each house surrounded by a piece of ground of which they make a garden. If you will go through any of these, you will find oranges, bananas, pine-apples, mangosteens, melons, shaddocks, and lots of other fruits growing there."

The settlement into which the little party had entered consisted of some fifty or more bamboo huts, one-half of which stood on one side of the regular highway and one half on the other. As each dwelling was placed several rods back from the road and was imbedded in vegetation, the scene was charming, as viewed by one walking along the street.

The natives were passing to and fro, and their pleasant voices were heard in every direction. It was the regular dinner hour, for the Javanese as a rule eat only twice a day—once just before noon and again at sunset, when their day's labor is ended; but the excitement caused by the frenzied native had brought a good many upon the street, and they were talking loudly about the startling occurrence.

Jurak paused long enough to inquire of some of his friends and to receive their replies. Then, as they resumed their walk, he explained:

"He was a good man, but he took too much opium; he became crazy a year ago and was frightened, so he stopped for a long while; but he began to use the evil stuff again. I was talking with him this afternoon and saw the wild look in his eyes. I suspected what was the matter and stayed by him; but when my head was turned, he gave a great shout and started to run amuck."

"Did he hurt any one?" asked Eustace.

"He cut several people, but not very bad; he knew I was at his heels and he was afraid of me, so he had not time to stop very long on the road. I thought when he turned toward you two, when he met you, that you would both be killed."

"*I didn't*," said Hermon, "unless my gun should miss fire, and it has never done that."

CHAPTER XV.

A JAVAN HOUSEHOLD.

JURAK glanced admiringly at the youth, whom he held in high estimation for his exploit in killing the tiger.

"You would not have been blamed if you had shot him, but I am glad it did not become necessary."

"So am I; indeed it would have been a sad thing even in self-defence. How is he now?"

"He is in his hut off there to the left, where his wife and friends are taking good care of him. They have a native physician who has given him medicine that will soon put him to sleep. Here is the hut where we will stay to-night."

Turning abruptly to the right, Jurak led the way through the charming garden which inclosed one of the ordinary bamboo huts of which mention has been made. There were no fences surrounding these yards or gardens, and they walked over a path so thickly strewn with fine gravel that it seemed to Hermon he had never made such a racket with his feet, which would have soon become tired had he walked further.

The inviting appearance of everything became more marked because of a long bamboo bench in front of the structure and a little to one side. On this sat a native smoking a long-stemmed pipe, with the contented stolidity of a Dutch burgher on the banks of the Zuyder Zee.

As he saw the little party coming up the graveled walk toward him he rose to his feet, took his pipe from his mouth, and held it to one side while the other hand was thrust part way forward as he smilingly awaited an introduction.

This ceremony was performed by Jurak very much as Eustace himself would have done had the situation been reversed. The name of their host was given as Dati, which you will recall is the Javanese word meaning Sunday.

Dati was similar in dress and appearance to the majority of the natives whom Hermon had seen since his arrival in that country. His long black hair hung unconfined about his shoulders, though he sometimes wore it plaited, with a handkerchief bound around his crown in the form of a turban. While many of the people go barefooted the year round in Java, both Dati and Jurak wore shoes and stockings such as are seen in this country. Each had the long petticoat-like garment peculiar to the island which begins either at the waist or shoulders and descends to the feet, being clasped at the waist by a girdle, in which is generally carried the *kris* or formidable knife of the country. Our two friends were covered by the long, flapping garment described, and underneath it was a pair of trowsers, with a short vest inclosing the upper part of the body. I should have said that when Jurak was first seen his head-gear was a large hat of palm leaves in the shape of a reversed wash-basin, which he flung aside as he took his seat on the bench alongside his friend, who also made room for the boys. Dati set the guns inside his hut and then invited them to the rear, where they bathed their hands and faces, took a swallow of clear, pure water, and respectfully saluted the wife of Dati, who was busy preparing supper for them. Like most of her sex in that country she was very homely in appearance, and both she and her husband had the stained teeth that are so shocking to look upon.

But she was an excellent woman and a good wife, who treated her guests with great kindness. What particularly interested the cousins, however, was a little boy about six years old, who was the only child of the couple. He was fat, jolly and pretty. A small piece of colored calico bound around his loins was all the clothing he wore. As a rule, the native children of Java wear no clothing at all until seven or eight years old, and spend nearly all their time outdoors, until perhaps twelve or fourteen years of age; then they must help their parents work.

The little fellow, who was called Woo Woo, was shy at first; but he soon gave the visitors his confidence, and they made a great deal of him. He could not speak English, though he had no difficulty in expressing himself in his own tongue. As the mother was equally ignorant, Eustace acted as interpreter while the husband was in front engaged with Jurak. Dati could talk a little in English, but very brokenly.

The supper (or dinner) was eaten in the usual fashion. A handsomely-colored mat was spread on the ground and the party seated themselves cross-legged around it. Hermon noticed what he had observed in the house of Jurak, that the rice with which he was furnished was colored; but instead of being yellow it had a fine brown tint, due, as in the former case, to some coloring matter which the housewife had added in cooking. The eggs are tinted, as we are accustomed to see them during Easter-time; and even pies and cakes are always colored.

To Hermon it was a curious feature of life in Java that, though it is the granary of the East Indies and one of the great coffee-producing districts of the world, yet the berry itself is used very sparingly in that country. When the cousins sat around the board of Jurak there was no coffee seen, nor did Dati bring forward any. In its place he offered boiled water, which the lads gracefully declined, confining themselves to the liquid in a cool form.

Rice is the chief article of food among the Javanese, though they have a great variety of fish, flesh, fowl, and vegetable. As they are Mussulmans, of course they never touch pork; but they eat almost every other kind of meat, and are generally addicted to hippophagy. Of course you know what that is.

As the boys had eaten nothing but fruit through the day, they were in form to enjoy something more substantial. Not only did they partake of rice, but also of fried buffalo steaks, besides several eggs, which had been previously packed in salt and ashes until the salt had penetrated the shell and flavored the egg itself.

There were two dishes on the table that the boys did not touch: one was composed of a species of worm found in the teak and other woods, while the second was made up of white ants, which had been caught in large basins and then drowned.

It was not yet dark when the meal was finished and the guests moved around to the front of the building, where Jurak and Dati relit their pipes and began a conversation in which both seemed deeply interested. Eustace occasionally listened and asked or answered a question, but when his cousin inquired what it signified, he laughed and replied that it was of no special account.

But while the men smoked the youths devoted most of their attention to Woo Woo, the bright little fellow, who showed great fondness for them. They rolled him about on the grass and frolicked with him until his merry laughter attracted the attention of those who were walking along the highway. The boys gave him a number of bright new coins, which delighted him simply because they looked pretty; but the parents, who glanced slyly at the pieces, knew their value and could hardly conceal their pleasure. The Javanese receive but the most meager wages from their Dutch employers, and the money which the boys gave to young Woo Woo amounted to more than his father could earn by a week of hard work. No wonder, therefore, that the parents were pleased.

Pretty soon Eustace left Hermon alone with Woo Woo, and sat down on the bench between Jurak and Dati, to whom he talked several minutes in their native tongue.

"I should like to know in what you are so interested," remarked Hermon, when his friend rejoined him. "Where's he going?"

This question was caused by Dati, who rose abruptly to his feet, said something to Jurak, and then walked out to the street, where he was quickly lost to sight in the gathering darkness.

"How should I know?" asked Eustace in turn, looking around as though the departure of their host was a surprise to him, when in truth he had been urging his departure.

"He acts as though he was going down to the corner grocery to talk politics or buy something for his wife," added Eustace in a careless manner, as if the question possessed little interest for him.

"It looks to me, from the way in which you talked to him, that he had started off to get a little rest!"

"It may be," was the meek reply of Eustace.

But only a few minutes passed when Dati reappeared, walking quite fast. Before he could resume his seat Jurak and Eustace were plying him with questions, and Hermon felt piqued that his cousin should refuse to let him know what the subject was which possessed such interest to both.

"I sha'n't question him any more," said the younger to himself. "If he chooses to affect so much mystery, I won't please him by showing that I care anything about it—heigho!"

A new arrival appeared. A native of medium stature, clothed in rich and brightly-colored costume, with hair elaborately plaited and coiled about his crown, walked deliberately up the path and approached the little party in front of the dwelling. Hermon at once suspected the truth; he was a distinguished native—a prince, in fact; the ruler of the district.

Before he reached our friends Jurak and Dati rose to their feet, took several steps toward him, and then both squatted on their heels, where they sat during the curious interview. This fashion, which struck Hermon as odd, is the Javanese method of showing respect for an eminent personage. To stand erect at such a time, even with uncovered head, shows ill-breeding.

Eustace and Hermon, however, were not expected to comply with the custom, since they belonged to a superior race. They looked curiously on, the younger especially wondering what it all meant.

The prince talked several minutes with the two men squatted on the ground, during which, if Hermon had looked at his cousin, he would have seen him smiling, as if anticipating some sport at hand.

The conversation had continued but a few minutes when the prince, to the amazement of Hermon, turned about and walked toward him with his right hand extended.

"Shake with him," said Eustace in a low voice.

"Of course I will," replied Hermon, grasping the offered palm and cordially pressing it.

"Great man," said the prince, speaking brokenly and with a most expansive grin; "he great hunter—he shoot tiger—he great hunter."

It all flashed upon Hermon. His fame had reached the village. His friends had laid the plan of telling the ruler that the wonderful hunter had arrived, and he had called to pay his respects. Before the lad could frame any appropriate reply to the words of the prince he bowed low, turned about, and walked slowly down the graveled path. Jurak and Dati rose from their humble posture and resumed their seats, both joining Eustace in hearty laughter over the embarrassment of Hermon.

"That beats everything!" he exclaimed. "I know who set that job up on me, and I will get even with him."

"You have been greatly honored," said his cousin; "it is not every verdant American who can come to Java and be called upon by a native prince as a matter of respect."

"Why didn't he invite me to spend a few days with him? I suppose I ought to be grateful, but all the same I am not."

However, the best of feeling prevailed, and the party sat a long time on the bench in front of the hut talking together, though one of the number was somewhat handicapped, because he did not understand the language of the others. Finally, they retired to rest.

Dati and his wife and little boy occupied the rear apartment, while the others slept on the rude beds raised a few inches from the floor and covered with mats, which were spread in the front part of the structure.

Jurak and Eustace soon sank into a restful slumber, but it was a full hour before Hermon's senses departed from him. Then he would have slept until morning had he not been awakened in an unusual way.

At first he thought he was in a hammock on board ship. His couch was rocking to and fro in the most alarming manner. Hastily rising to a sitting position, he stared around in the darkness. The rocking continued, and was accompanied by a peculiar jarring and rumbling, such as he had never felt or heard before.

All at once the truth burst upon him: it was an earthquake!

CHAPTER XVI.

A MEMORABLE NIGHT.

WE ARE so accustomed, except in the latitude of Charleston, to look upon the earth as solid and immovable, that the shock caused by an earthquake may be said to be greater morally than physically. Hermon Hadley lay still only a moment or two after learning the frightful truth. Grasping the shoulder of Eustace he exclaimed:

"Quick! get up! don't you feel the earthquake?"

He was obliged to shake his cousin several times and to repeat his alarming words before he aroused him to a sense of the situation. Eustace was very sleepy, and when he assumed a sitting position muttered:

"It *does* seem to be an earthquake—that's so; but one place is about as good as another. This house won't tumble down, and if it does, it won't hurt anybody."

"Let's get outdoors!" exclaimed the younger, who could not shake off his terror and was hastily dressing himself. It took but a short time to don his clothes, when he ran outside, followed a minute later by Eustace.

Though volcanic eruptions and earthquakes may be said to be a feature of every-day life in Java, yet no one can become so used to them that he does not feel a terror when the earth trembles and sways beneath his feet. The earthquake which awakened Hermon Hadley was severe at first. The ground rocked violently back and forth, then stopped, trembled slightly, and just as he reached the outside of the hut there came such a wrenching lurch that he staggered and narrowly missed falling.

"My gracious!" he exclaimed; "what will become of us?"

"This *is* rather rough; fact is it beats any I can call to mind," remarked Eustace, who stepped outside a moment later still adjusting his clothing.

At such times animals no less than men are sensible of the appalling convulsion of nature that is going on. Jurak, Dati, and even Woo Woo and his mother came hurrying forth and grouped themselves in the moonlight outside. The villagers were running back and forth along the street, or standing in groups by the wayside, talking together in low tones, and tremblingly awaiting the next shock. Others were kneeling in prayer; while numbers of dogs, including Tweak, who, like a worthless coin, turned up when not wanted, were galloping back and forth, some barking, and others moaning and whining in terror.

Jurak, Dati, and the wife of the latter, stood

motionless, side by side, talking in low tones and glancing at the clear, moonlit sky, and occasionally exchanging a word or two with people who were passing.

After the dizzying lurch made by the earth the vibration became of a gentle, tremulous character, such as would have done nothing more than cause the rattle of the window-panes, had there been any in the houses; but everyone was expecting another shock, and the expectation of such a thing is almost as terrifying as the occurrence itself.

"I tell you," said the white-faced Hermon, catching the arm of his cousin, "I would rather live in the worst part of my own country than in the finest section of the world and have such shocks as this!"

"I think you have earthquakes there, too; I know they do in California."

"But they are nothing like this—only slight jars, which the people don't mind. Why, this earthquake was enough to shake down the whole city of San Francisco!"

(You will bear in mind that these incidents took place before the recent seismic disturbances in Charleston and its vicinity.)

"Or to turn the buildings around, so as to make them face the other way," remarked Eustace with a laugh which amazed Hermon; for of all times for mirth, this was surely the most inappropriate.

The jarring rumbling ceased, and all became as quiet as before. Men and women began strolling back and forth, as though it was early in the evening. They talked in louder voices, and a number came up the graveled walk of Dati and held converse with him and Jurak. Woo Woo and his mother went inside and lay down again, as though they considered all danger over.

"I guess we may as well do the same," remarked Eustace, who had sat on the bench in front of the hut and was yawning like a boy who ought to have gone to bed long before.

"Can you sleep at such a time!" asked the astonished Hermon.

"If it doesn't come too hard, what's to hinder? The houses in Java can stand a good deal of shaking before they go down, and these bamboo huts are the best."

"But sometimes the ground opens and swallows people."

"Yes," said Eustace, rising to his feet and passing into the hut, "that *does* sometimes happen, though I've never seen it in Java. If I had, probably I wouldn't be here to give the particulars. But come in and lie down. You needn't take off your clothes, and you're so well scared now that you ain't likely to sleep very soundly."

"Sleep!" exclaimed Hermon. "As though I could close my eyes at such a time! There may be a good many things I can get used to, but an earthquake is not one of them."

"Well, take it philosophically; that's the best plan."

"Do you think there will be any more shocks?"

"You know it must all be guesswork on the part of any one; but I've passed through a good many earthquakes in Java, and I shouldn't be surprised if we get another dose before the thing is over. However, I'll be obliged if you don't wake me up. If it should turn things topsy-turvy, I'll wake any way; and if it don't do that, a fellow needn't bother."

Eustace had lain down again, and he now turned over on his side with his back toward his cousin.

"Well, if that doesn't beat everything!" remarked Hermon, carefully reclining beside him. "I never would have believed that any one could be so indifferent. I'm sure I never could learn to feel as he does."

A half hour passed and no more shocks came. Then Jurak and Dati entered and lay down, and by and by all became still. The villagers seemed to have gone to sleep once more, and Hermon began to hope that all danger was over. Just before sinking into a restless sleep, however, he was startled by a slight jarring, which soon ended.

CHAPTER XVII.

JOURNEYING SOUTHWARD.

IT WAS not until the night was nearly past that Hermon Hadley sank into a deep, refreshing sleep. The earth had settled to rest, and so when slumber did overtake him it lasted longer than usual. When he opened his eyes the sun had been above the horizon more than an hour. Everyone who had slept within the hut was outdoors, and the mischievous little Woo Woo had pulled the young American's nose, as though he thought it too short and wished to lengthen it.

"What are you at, you little rascal?" demanded Hermon, catching the youngster, who yelled and kicked and laughed, and tried to wriggle himself loose. The lad gave him a gentle spanking and allowed him to get away, and then, adjusting his clothing, passed outside, where the morning meal, as was the custom on the first day of the week, was awaiting them.

Hermon found not only the friends of the night before, but Myeta, the wife of Jurak, was there, and greeted him pleasantly. The day was the Sabbath, and she had come to spend it with her husband among their friends. She still looked upon Hermon with great admiration, for was he not the hunter who had killed the terrible tiger?

All were talking about the earthquake, but, so far as could be learned, it had caused no particular damage. As it was Sunday, Eustace and Hermon concluded to stay where they were until the morrow, Jurak having promised them that if they would do so he would accompany them on their visit to the Sacred Mountain. It did not seem right to the lads that they should spend the holy day in wandering for pleasure over the country, and it was therefore to their credit that their rest was more a matter of conscience than of inclination.

But there was nothing to be expected in the way of religious exercises in the village—that is of a Christian nature, the boys being the only two beings in the place that bore the name. Although, as I have told you, the Javanese are Mohammedans, yet engrafted upon their religion are fragments of all the creeds with which they have been brought in contact during the centuries of their history; so that the result is a mixture beyond the power of any man to systematize. The number of spirits worshiped by the Javanese is past computation. They are everywhere, and the smallest hamlet on the island has its own special saint, who is the author of the good and bad that come to the people.

Near the center of the village in which our young friends were spending the day they observed a great, spreading tree (most likely the direct cause of the village being located around it), beneath which was erected an altar. The natives were continually approaching this from morning till night. As they did so they laid on it offerings of incense and flowers, uttering in broken Arabic the old formula, "There is no God but God, and Mohammed is his prophet."

There were other ceremonies—such as the repetition of prayers in their houses—which continued until after the sun went down, and which were the cause of serious thought and meditation on the part of the boys. All this was especially impressive to Hermon, who witnessed it for the first time, and who recalled the missionary exercises in which he had engaged with his own Sabbath-school, so many thousands of miles away.

It being the first day of the week, three meals were eaten—that of the morning, however, being quite light. The boys took several strolls through the village, and though they attracted considerable attention it never was of an unpleasant character. The day, however, became monotonous, and they were glad enough when the night which followed was ended; and bright and early the next morning the two, in company with Jurak, and with Tweak circling back and forth through the woods around them, started southward on their way to the Sacred

Mountain, which they expected to reach within a couple of days.

Less than an hour had passed when, on the crest of the mountain chain which extends east and west through the whole length of Java, they were favored with the most magnificent view on which the eyes of Hermon had ever rested. To the right and left stretched the vast slopes of the mountains, emerald with vegetation, and dotted here and there with towns and villages, some of them so smothered in foliage that the eye could not detect them. There were winding streams, uniting further on, and expanding as they neared the sea; there were vast open spaces, covered with the long silvery grass peculiar to the country; piles of black, volcanic formation; towering peaks, several of the highest climbing away up in the clouds to the snow line, and from the summits of more than one of them poured out enormous masses of inky vapor, which floated off horizontally and seemed to settle in motionless clouds along the sky. When amid the murky discharge, as it rose above the crater, the inhabitants saw red tongues of fire darting back and forth, they trembled, for they might well dread the consequences.

But shading their eyes and gazing intently southward, the boys could make out a blue, fleecy line stretching around the horizon, until it gradually vanished and mingled with the sky. This faint bluish line was the sea, and had they possessed only an ordinary spy-glass they could have detected more than one snowy sail stealing across its vast bosom.

"And that is the Indian Ocean," muttered Hermon Hadley, his eyes roaming from east to west over the whole extent of the line; "the smallest of the five oceans, but still too vast for any one to comprehend its immensity."

"The coast on that side of the island," said Eustace, "is high and rocky, while on the north you know it is low and marshy."

"And unhealthy."

"No doubt of that. Batavia has been called the graveyard of Europeans, for many have met their death there. But the foreigners have located further back, on higher ground, where they find the climate much better. Along that coast," added the elder, pointing to the south, "are gathered one of the peculiar products of this country."

"What's that?"

"Edible birds' nests. They are those of the swallows, which build them among the rocks and caverns. Thousands are gathered by the natives and sold, for you know they are considered great delicacies—the wealthy Chinese esteeming them highly and paying large prices for them."

CHAPTER XVIII.

AFTO.

JURAK was too much accustomed to the beautiful scenery of his native country to feel the enthusiasm of his young friend from the other side of the world. He stood patiently until the boys were through enjoying the magnificent prospect, when he resumed the walk toward the south.

The party were following one of the regular highways which led down the mountain slope toward the ocean, and over which many people beside themselves were traveling. Numbers were on foot, others were riding the tough little horses of the country, while still others were seated in carts drawn by horses or oxen or buffaloes.

In every case the parties whom they met saluted Jurak and the boys, and now and then exchanged a few words with the native. Each one was respectful; but there was something in the appearance, manner and lives of these Javanese which painfully impressed Hermon Hadley. While they were courteous, obedient and law-abiding, they were thus, it seemed, because they were crushed to the earth by their Dutch masters. More than once, as the young American remarked to his cousin, it appeared to him that Java was a great prison where all the inhabitants were allowed to walk about and work outdoors so long as they behaved themselves.

"It is best that it is so," said Eustace with a laugh, "for these people in ancient times were as warlike and cruel as the red Indians in your own country."

Jurak informed them that in order to reach the Sacred Mountain by the shortest route it was necessary to leave the well-traveled road over which they were walking and bear to the west. Many of the ruins in Java are in the depths of the forests, where they cause so little interest that the visitors are not numerous enough for their feet to leave any impressions in the moist earth surrounding them.

It was yet comparatively early in the day when the three turned off into a branch road which, by the faint imprints, showed that it was little traveled. It should have been termed a path, indeed, for it was so narrow that no wheeled vehicles ever passed over it, though the ground was well worn. Jurak explained that it was a route taken by many of the natives, scattered over that section of the country, who journeyed on foot or horseback and were seeking the main road, which the party had left a short time before.

They had gone only a little ways along this path

when Tweak was seen running from the opposite direction as though coming from a distant point; but as he left the village that morning, it was clear he had been frolicking around them for the entire distance. Without coming very near, he wheeled about and trotted out of sight around a curve some distance ahead.

The country through which they were journeying was similar in many respects to that over which they had traveled after leaving the hut of Jurak, on the other side the mountains. Luxuriant vegetation inclosed them at all times. Sometimes it was in the form of spreading trees, whose limbs were so interlocked that the wonder was how they could grow at all; and again it was the tall, growing grass, sprouting with such vigor that it almost hid from sight the innumerable masses of black-colored stone.

Some two or three miles further on our friends met a man whose resemblance to Jurak was so striking that the cousins instantly remarked it. He was of the same height, nearly the same age, and dressed very like him—that is, he wore good European shoes, with the long petticoat, the bottom of which was looped up, as is the fashion among the natives when at work. The sash around his waist held the long *kris* or knife peculiar to Java, and his head was surmounted by the basin-like hat similar to Jurak's. He had no fire-arms, however, his only other weapon consisting of a long, sharply-pointed spear--

an instrument which the natives handle with great effectiveness and skill.

It was evident from the manner of Jurak and this man when they first saw each other that they were old acquaintances. They smiled, shook hands, and both talked very fast at the same time, so that Eustace laughed at the exhibition.

The animated conversation continued only a few minutes when Afto, as he was called, turned about and walked with the party along the path in a south-western direction.

"Is he going to make one of our company?" asked Hermon of his cousin, who had listened closely to the conversation.

"Only for a few miles. He says that a couple of fierce leopards, which have destroyed a good many animals in this neighborhood, committed so many depredations last night that he was on his way to get help to hunt them down."

"Where does he live?"

Before replying, Eustace held some conversation with Afto. When through, he said to his cousin:

"He is a laborer on a coffee plantation a few miles further down the mountain. He lives with his father and mother, nearer the plantation than any of the others, but there are plenty of them scattered around the neighborhood. They have fowls, goats, sheep, and larger animals. Some months ago a number of them were killed by some wild beast. A good many thought it was a tiger, for the tracks were very much like those made by that beast; but those who were more experienced said it was not a tiger. One moonlight night several of the natives saw him, and then it became known that it was a black leopard—which is almost as much dreaded as the tiger. After awhile it was learned that his mate was traveling with him, and then, you may be sure, there was dismay through the country. The people organized parties to hunt them; but the leopards kept out of the way, though they came near cornering one of them. When the folks were becoming desperate, and gathered in such large numbers that there seemed no escape for the beasts, all signs of them disappeared. They had gone so far into the mountains that it was not thought worth any more attention. But now and then one of them would make a visit, kill some animals, and be off before the people could gather to attack him."

"Has he hurt any people?"

"I will ask Afto."

CHAPTER XIX.

THE POST OF HONOR.

EUSTACE was surprised, on questioning the native, to learn that two men had been killed by the leopards, and that within the preceding week a little child had been borne away from in front of its home and never was seen again.

"Why does he not go on and arouse the country, if there are not enough people in his own neighborhood to slay the animals?"

"He would have done so had it not been for Jurak, who told him that we three knew how to use our guns on wild animals. When he said you had killed a great tiger, that settled the question with Afto. You are the young gentleman that's wanted; they are waiting for you; they will not be happy until you hurry to their relief."

The expression of disgust on the face of Hermon became so pronounced that his cousin laughed heartily. He could not doubt, however, that Eustace had told very little more than the truth. The admiring looks which Afto cast upon him could have been caused by nothing except a knowledge of Hermon's exploit, and he must have felt that such a formidable hunter would have little trouble in disposing of two leopards, even though they were exceptionally large and bold.

Afto reported, further, that both the animals, although not corraled, were known to be in a thicket or jungle not far away. The plan was to leave them undisturbed until enough hunters could be gathered to surround the place; then they would close in and finish them.

"Is Afto the only one who has gone out to gather recruits?"

"Two others are engaged in doing the same thing. They have taken another direction, and will not go far. The fact is, as I am certain from what Jurak said, that they have got enough together to kill both beasts; but the people are in a state of panic and hesitate to make the attack. One thing is certain—no one can be more welcome than we."

"They must have fire-arms—that is, some of them."

"Yes, they have all sorts of weapons; but the fire-arms are not very plenty, and none of them are as good as ours. It takes a long time for all the modern improvements to reach Java—halloo! Tweak must have started up the leopard!"

The dog came galloping back in great terror. His ears were laid flat, his tail was down, and he kept glancing backward as if some furious beast was at his heels.

It was not impossible that one or both the animals had been disturbed by the canine, and Jurak and the boys held their guns ready to use at an instant's notice, while Afto grasped his spear with a firmness and vigor which boded ill for any foe that might cross his path.

All at once the four broke into loud laughter. One of the wild pigs that are so numerous in Java came trotting straight after Tweak. The porker was alone, and now and then he gave a little grunt, as if enjoying the sport. Catching sight of the hunters, he threw up his head, looked wonderingly at them a moment, and then, as if satisfied, he emitted another grunt and turned off into the woods.

It would have been an easy matter to kill him, but he was hardly worth the powder; and with many slurs upon the dog, who had shown similar cowardice before, the party pushed on toward the thicket, which Afto informed them was not far off.

Hardly a quarter of a mile was passed when they debouched into an open space covered with grass and several acres in extent, on the further side of which stretched a dense piece of matted woods. While this reached right and left until it joined larger bodies of forest, it was no more than a hundred yards in depth; that is, if our friends had pressed straight on they would have gone clear through the neck and emerged into the coffee plantation on the other side.

Fully fifty people were scattered along the woods. Several had rusty old firelocks, but the majority held spears in their hands, besides the knives in their girdles. They were not shouting or making any demonstrations, but were walking idly back and forth and talking in low tones, as if afraid to make any demonstration against the enemy before other reinforcements arrived.

Such was the fact; and the appearance of Jurak, Afto, and the two tall youths with their rifles, was most welcome indeed.

"For gracious sake don't let Afto tell them anything about the tiger!" said Hermon, as the parties mingled.

"I'll speak to him," replied his cousin, stepping hastily forward and touching the arm of their new acquaintance. The latter turned and listened to the earnest words of Eustace. Then he nodded, walked away, and immediately began talking and gesticulating very fast to his friends, who one after another turned their heads and gazed with such marked wonder and admiration upon Hermon that he became suspicious and wheeled around to Eustace.

"What in the mischief did you tell him?"

"Didn't you ask me to urge Afto to give the particulars of your killing the tiger, so as to shut off Jurak before he could draw a long-bow?"

Eustace looked so serious that Hermon was deceived for a moment; but he detected the twinkle in his cousin's eyes and observed a twitching at the corners of his mouth. Then he made a plunge for him, and the taller youth had to do some lively running and dodging to escape the wrath of the other.

"Well, if they *are* resolved to make a hero of me," remarked Hermon, forced to stop and laugh at his mischievous relative, "I can't help it; but I'll even matters with you, young man, and don't you make any mistake."

Meanwhile the natives were discussing the situation. The boys joined them, and Eustace caught every word. In a few minutes he turned about with sparkling eyes.

"What do you think, Hermon? They have agreed that as you and I have such excellent guns we are to take the lead; and inasmuch as you killed the tiger, you shall have the first chance to slay the leopards."

"Good! Nothing can suit me better," said the plucky young American. "My gun is loaded and I am ready. I hope they won't keep me waiting."

"No fear of that," was the reply of Eustace.

The presence of Jurak, Eustace and Hermon Hadley, each with his loaded rifle, gave to the natives a confidence which they had lacked up to that moment. A brief conference was enough for them to decide upon the plan by which the leopards were to be slain. There were fifty men on one side of the neck of forest and about the same number on the other side, while between, the two dreaded animals were believed to be crouching and awaiting attack.

For these parties to communicate with each other, it was necessary to pass through the narrow woods which separated them and in which the game was hiding. They made the passage by diverging far to one side of the dangerous spot.

The plan of assault was simple, being mainly that of surrounding the beasts. This compelled a line to extend through and across the neck of forest, one on the right and the other on the left. These were to advance simultaneously with those from the edge of the clearing. Thus the ring would gradually narrow and become more compact as it neared the game.

It will be seen that it was beyond the power of any one to determine beforehand the most dangerous point in this contracting circle, for the precise whereabouts of the animals were unknown; but Eustace, Jurak and Hermon were stationed on opposite sides and well in advance of the natives who meant to back them up. Those who had firearms were closer, while the spearmen formed what might be called the rear-guard, and calculated on being "in at the death." Tweak had vanished long before and none of the natives had any dogs, for they were of no account at such a time.

Some of the Javans were still absent in quest of reinforcements, but the arrival of our friends rendered it unnecessary to await them. Within fifteen minutes of the time when the little party emerged into the clearing the large circle of hunters began closing in upon the game.

The diameter of the huge ring when the advance was begun was nearly two hundred yards, and as there was considerable undergrowth, the opposite sides were not in sight of each other. In fact, in some instances a native could not see his neighbor on the right or left; but as the advance steadily continued they necessarily approached, and soon the swarthy faces were gleaming almost everywhere through the woods.

It need not be said that Eustace and Hermon, while alive to the honor done them, realized the dangerous business they had entered upon. Being in advance of the others, they were the most likely to be assailed by the leopards, which, when driven to the wall, will fight with the fierceness of tigers.

After all, the cousins dreaded the danger from the men themselves more than that from the beasts. There was likely to be much excitement and confusion, and wild shooting was sure to take place.

"I hope it will fall to the lot of Hermon to bring down one of the leopards," said Eustace to himself, as with cocked rifle he stole noiselessly forward, looking sharply not only to the right and left but among the limbs of the trees, for the leopard differs from the tiger in being a skillful climber. "He don't fancy being made a hero, but it will be something to boast of when he goes back home. There isn't many boys of his age that can truthfully say they have killed a leopard and tiger."

The natives had now advanced so far that they could catch glimpses of each other. Many of them were shouting and beating the bushes, with the hope of frightening the animals into darting from cover. All at once Eustace heard a sudden commotion directly opposite him, and he knew that one or both animals had been driven out. Men were running together, shrieking and leaping about, brandishing spears and firing their rusty muskets, until the commotion was enough to make one believe a dozen tigers had assailed them.

Only one of the beasts, however, had appeared, and the assistance of others was not needed. Although sharing in the excitement they stayed away, using eyes and ears to the utmost to guard against an attack from the other, which, it was natural to suppose, would rush to the help of its imperiled mate. But the leopard is cowardly and sneaking until forced to fight; then, as I have stated, he becomes a tiger in his fierceness. Although the female was assailed on every hand, and emitted snarling shrieks of rage and pain, the male kept away from her.

The female proved she was "game." Finding herself unearthed from where she was crouching under a clump of bushes, she glided noiselessly forth and stole swiftly toward the dusky native who was shouting and beating the shrubbery with his spear. The latter quickly espied his danger, and with a cry which nstantly brought his friends to his assistance he turned and ran for life; but fast as he was, the leopard was faster. With a single tremendous bound she rose in air, and shooting through the limbs which intervened, dropped fairly on the shoulders of the fugitive and bore him to the earth.

At that moment the nearest hunter drove his spear into the shoulder of the leopard and literally forced her off the prostrate body, which was already torn by her needle-like claws. The latter, infuriated by the attack, sprang for her assailant, who, finding his spear useless, drew his *kris* and struck viciously at the brute, which, however, persisted in her attack. The native was forced backward and dropped on one knee. The wound inflicted with his knife only added to the fury of the brute, which with inconceivable quickness wounded him grievously in return.

The fellow who was making such a brave fight for his life would have succumbed had not assistance reached him almost immediately. As it was he lay still, while a dozen men jammed their spears into the leopard, which, unable to withstand such an overwhelming attack, snarled and fought with unsmotherable fury for a minute or two only, when she succumbed to a score of wounds, any one of which would have proved fatal.

CHAPTER XX.

A SHOT IN TIME.

MEANWHILE events became lively in another quarter. Hermon Hadley, when the hunt began, felt doubtful whether either one of the leopards would be found where it was supposed to be. It seemed unnatural that after they had showed so much cunning and shrewdness, they should squat down and wait quietly while such formidable preparations were going on for their destruction. But the youth was too wise to neglect any precaution because of these doubts.

When the fierce wrangling announced the death-struggle, he stopped and instantly changed his mind.

"The other can't be far away," was his conclusion, halting by the side of a tall slim tree, "and as they seem to be able to manage one beast, we ought not to have much trouble with the other."

Seeing the "mighty hunter" stop, the natives near him did the same, leaving their friends undisturbed while they finished the first brute. Knowing the skill of the leopard in climbing, Hermon closely scrutinized the limbs in sight; and it was well he did so. The cries of the expiring female were still in the air when he detected a movement among the branches a short distance in advance and some twenty feet above his head. There was so much vegetation that he could not see clearly, but he was convinced it was the beast for which they were hunting.

The snarling of his mate in her last struggle did not bring him to her defense, but warned him that he was in a dangerous dilemma himself. The brute was among the limbs in search of some game for their dinner; but he gave over all design in that direction and moved out on a large limb, where he crouched and awaited developments. Rather curiously, the outcries and beating of the bushes caused him no disturbance. The female was frightened, however, into running into destruction, and then the male looked about to decide what was best for him to do.

The first object which seemed to attract his attention was a sturdy American boy, standing with gun in hand and looking intently upward at him. Without pretending to gauge the meditations of the leopard, it is safe to believe he felt little fear of that young gentleman; and although the sounds indicated there were many other enemies in the immediate neighborhood, it looked as if he believed he had only to drive that particular lad from his path in order to open the way of escape.

Hermon identified the animal when it leaped as lightly as a cat to a limb nearer him, and crawling swiftly along it, stopped, crouched again, and glared and snarled in a way that left no doubt of his intention to open hostilities.

"It strikes me that this is a good time to try a shot," was the thought of the young American as he brought his weapon to his shoulder.

The natives, who had stopped when he did, saw him level his weapon at something among the limbs, but they were not close enough to discern the object. No doubt, however, could remain as to its identity, and they were chivalrous enough to keep back and not seek to interfere with his amusement.

Hermon did not want any interference. The angry beast was close to the limb, pressing its sharp claws in the bark after the manner of a cat, and concentrating its muscles for a terrific bound at the youth.

At the very moment he was in the act of leaping Hermon Hadley fired, sending the bullet into the neck and shoulder of the leopard instead of through his heart, as would have been the case had the animal remained stationary only a second longer.

Knowing he would bound for him, Hermon jumped to one side at the instant, as may almost be said, that the ball left his gun. As he did so he caught a glimpse of the black creature, elongated by its effort, shooting downward as if fired from a gun.

He saw the glare of the eyes, the gleam of the teeth, and heard the fierce growl as it struck the ground like a log. It was up again and made for Hermon, who had retreated still further, and dropping his gun, drew his revolver and began shooting at his foe.

But the shot from the rifle was fatal, and after several blind efforts to keep its feet and to get at the lad it made a tremendous bound upward, emitting a snarl, and falling on his side, gave one or two convulsive movements and lay still in death. It perished so quickly, indeed, that when the other natives rushed forward all life was gone. The two leopards had been killed within a few minutes of each other, and, as Eustace Hadley desired, the fatal shot was fired by his cousin Hermon—the sturdy youth who had come all the way from the other side of the globe, and whose first performance was the slaying of the dreaded tiger.

His last exploit naturally raised him very high in the estimation of the natives; but while they gave him full credit for his bravery, coolness and skill, the majority were sensible enough to attribute much of his success to the excellence of the weapon he carried. When Hermon learned that such was the case, he encouraged the belief as well as he could by means of his interpreters, who, I am sorry to say, were not always reliable.

But the neighborhood was well rid of two danger-

As the leopard was in the act of leaping, Hermon Hadley fired.
(See page 179.)

ous pests, and the rejoicing was universal. Hermon was compelled to shake hands with his admirers over and over again. He conducted himself with becoming modesty, but when Eustace told him that it had been decided to hold a grand feast in his honor he refused. He declared that unless Eustace stood by him in the matter he would go forward alone to the Sacred Mountain.

Inasmuch as the elder youth also looked with disfavor on the scheme, the kind-hearted natives respected their wishes; but they promised that the skin of the male leopard should be given the young hunter who had slain it.

The two Javanese that had been wounded by the other brute were so badly hurt that it was necessary to carry them home, but it was generally believed they would recover, as happily they did in time. Afto, being among his friends and neighbors, bade Jurak, Eustace and Hermon good-by, and the little party pushed hopefully forward toward the Sacred Mountain.

CHAPTER XXI.

COFFEE.

ALTHOUGH Hermon Hadley knew that Java is one of the greatest coffee-growing districts in the world, and though he had taken passing glimpses, as they may be called, of the industry, yet he knew little of the method of cultivation. Now, as he and his two friends traveled for a considerable ways alongside the extensive coffee plantation, he studied the matter closely and asked many questions of his friends. Eustace answered fully whenever he could, and what little he did not know was told by Jurak, whose life had been mainly spent on the coffee plantations.

When a space has been selected for the cultivation of the coffee-berry, it is inclosed by a hedge of quick-growing plants placed a dozen feet beyond the outermost row of coffee-trees. Beyond this hedge a ditch is dug, so as to carry off the surplus moisture.

The plants are grown from the seed that are obtained in the nurseries. The berries, which remain on the trees until fully ripe, become dry and dark-colored. In that condition they are set out and lightly covered with soil. They remain until two leaves sprout forth, when they are transplanted into beds a foot apart and protected by sheds from the sun. A year and a half after the first transplanting they are ready for removal to the plantation, where they yield one of the most important products of commerce.

The coffee plantations are laid out in squares. Holes two feet in depth and six feet apart are dug in which the young coffee-plants are placed; and in the middle of the square, between every four plants, a shade-tree is planted to protect them from the fervid rays of the sun.

The young plants are very sensitive, and unless handled with great care are destroyed. The most elevated coffee plantations in Java use few of the shade-trees, but they cannot be dispensed with on the lowlands.

A noteworthy fact was told Hermon. In the lowlands the coffee-berry, although large, is almost tasteless; but the smaller ones, grown in the higher regions, possess an exquisite flavor.

The coffee-tree bears the second season after transplanting. At the close of the rainy season, during which the plantation is set out, the unthrifty plants are replaced, and thenceforward the principal work is to weed and cultivate the ground. No one ever prunes a coffee-tree, which **frequently grows to the** height of fifteen feet.

The fruit ripens during nine months in the year.

The gathering generally begins in June or July, and ends in March or April. Each pod or husk contains two grains of coffee, which are fit to gather as soon as the pods turn a deep red color. They are picked one by one, great care being taken not to injure the buds and blossoms scattered among the riper pods. A light bamboo ladder is used, and the women and children generally gather the crop, while the husband works elsewhere.

The average yield of a coffee-tree is about two and a half pounds, though ten times that amount has been gathered from one tree. There are three gatherings. The first is light; the second the most abundant, and the last is what may be called the aftermath, since it is the gleaning of the crop. In the lowlands of Java the yield continues for eight or ten years, and in the higher regions it is longer.

The newly-gathered berries are spread out on hurdles several feet above the floor of the drying houses. Beneath these hurdles a slow fire is kept burning day and night. The movable roof of the drying-house is taken off morning and evening so as to give the berries fresh air, but during the middle of the day it remains in place. Coffee when dried in the sun is lighter in color and weight and larger in size than when artificially dried; but the latter has much the finest aroma, and the Javans insist that only when coffee is dried by a wood fire does its flavor become perfect.

The berries being fully dried, they are placed in bags of buffalo-skins and pounded until the husk drops off and liberates the coffee-beans. These are then separated from the husks, placed in bags or baskets, and deposited on shelves or platforms some distance above the ground, where they stay until carried to the nearest sea-port, thence to be shipped to the four quarters of the globe.

By the time Hermon had absorbed all this information they were beyond sight of the coffee plantation which was the cause of the boy's interest in the question. The afternoon was not far along when they passed through a small village, where Jurak found several acquaintances with whom he exchanged greetings. At the invitation of one of them they partook of food, and then, after a brief rest, pushed on again.

They had advanced far, and although the elevation was much less they caught frequent glimpses of the Indian Ocean as it gleamed clear and blue, many miles away, against the sky. The view was extremely pleasing, as it always is when the sea is seen from afar.

Had they cared to advance more rapidly, a few hours would have brought them in sight of the Sacred Mountain; but, as I have shown, there was no call for haste, and they made up their minds to travel at a leisurely rate. When Jurak expressed some apprehension that the father of Eustace might not fancy one of his principal employes going off on such a long jaunt without permission, Eustace assured him that he would make everything right between them.

As was expected, Tweak turned up when the halt was made at the village and claimed his share of the meal. He ate so much that the natives expressed wonder how he could contain it all, and Hermon suggested that the canine be given their entertainers in payment for what he had eaten.

"I must admit he hasn't shone very brilliantly on this jaunt," replied Eustace, when his companion taunted him with the worthlessness of the dog, "but we haven't got home yet. Wait till we shall have done so, and then we will balance accounts."

"He has one good trait," said his cousin.

"What is that?"

"He doesn't oppress us with his presence, especially if there happens to be any danger. He is off again, and we may not see anything of him until to-morrow; but the affliction of it all is that we are sure to see him *some time*. If he would only make his absence last forever he would be a perfect dog."

CHAPTER XXII.

THE FIRST VIEW.

LEAVING the volcano of Sumbeng far to the westward and that of Merapi as far to the east, the little party advanced through a valley running directly south, and walled in by two mountain ridges extending in the same direction and gradually approaching each other until, within a few miles of the coast, they became one.

Between these mountains and somewhat closer to the western side winds the Ella River, which some twenty-odd miles from its outlet takes the name of the Praga, one of the principal streams of Java. The Sacred Mountain, which was their objective point, is within sight of the Praga.

The travelers were approaching the southern border of the district of Kadu, which adjoins that of Bagelen, in which is the Sacred Mountain. The latter district extends to the Indian Ocean, so that our friends were well across the island of Java.

Remembering how thickly populated the country is, it can be understood that no need existed for spending a single night outdoors; but when they found the sun low in the sky while they were making their way down the Ella River they decided to pass the night in the woods, reserving the luxury—if such it may be termed—of sleeping within doors until their return journey, when they expected to be in more of a hurry.

There was an abundance of the fruit of which I have often spoken all around them, so that they secured without trouble what they wanted. Numerous tropical birds were seen at intervals, and there was reason to believe that some of the more formidable wild animals of the country were not far off.

The party were near the settlements, for in the stillness of the sunset hour, when the trade-wind seemed to be at rest, the voices of persons were occasionally borne to them with startling distinctness. Indeed, before darkness had fully come they observed two men standing on the other side of the narrow river and looking at them with some curiosity. Being hailed by Jurak, one replied that they lived in Jokjokarta, which is a town lying nearly if not quite twenty miles to the south-east, in the district of the same name. After some further conversation the strangers withdrew.

Left to themselves, Jurak and the boys gathered a lot of fuel and started a rousing fire. The weather was so pleasant that it was not needed for comfort, but was done for the cheer it added to the occasion. Tweak had not yet put in an appearance, and no one cared whether he was seen again or not.

The fire was kindled where its reflection extended far out on the river and almost to the opposite shore. The moon, which had befriended them more than once before, did not rise till quite late, and it was intended that when it appeared, or soon after, they would lie down in slumber. They had done considerable tramping during the day and were quite tired.

Jurak filled his pipe, and reclining on the ground smoked slowly, with his eyes resting on the coals, and with that dreamy expression which showed his thoughts were far away. The boys did not venture to disturb him, but spreading their blankets on the ground they lolled about in the same lazy fashion and talked of everything that came into their minds.

"How about crocodiles?" asked Hermon.

"I don't believe any are found in the river, though there may be some. There are plenty of caymans in the lowlands, and I suppose a few make their way as high up as this. You aren't afraid of them, I hope?"

"Not that I know of. There are so many nuisances in Java in the form of birds, beasts and reptiles, that a fellow soon becomes accustomed to them. I only thought of crocodiles because I happened to notice the gleam of the water."

By and by the friends fell to talking of home and friends, until they became quite sentimental and homesick. Jurak took no part in their conversation, and when at last they laid their heads down and sank into refreshing, dreamless slumber, he was still gazing into the smouldering embers and slowly puffing at his pipe.

The boys were awakened by the crack of a rifle. Starting up, they found it was growing light, and Jurak was in the act of reloading his gun. He had just brought down a bird flying by, and in answer to the inquiries of his young friends said they would not only take breakfast at a civilized hour, but would partake of more nourishing food than fruit. As for himself, he missed his daily rice, and the only substitute that answered was flesh, which he had just arranged to furnish all.

The bird which he had shot was the size of a turkey, and despite the fact that it belonged to a tropical region, its feathers were almost as sober in color. It was sound and plump, and when broiled over the coals made an enjoyable and nourishing meal. The party were only fairly through when, to the amusement of all, Tweak walked solemnly forth, halted in front, wagged his tail and licked his chops.

His actions were so pleading that he was furnished with all he could eat. Then he frolicked around the camp for awhile until they resumed their jaunt. Finally he whisked out of sight, and nothing was seen of him again for a long time.

The three were in the best of spirits, and walked

at a brisk pace until the sun was high enough to feel its rays. It was not long before they struck a path which appeared familiar to Jurak, and quite frequently they encountered natives, who were invariably polite and pleasant.

It was not yet noon when, near the Praga, Jurak, who was slightly in advance, stopped and waited until his young friends stood by his side. Then with a suggestive smile he raised his bony arm and pointed to the south. Following the direction, they saw one of the spurs of the mountain ridge which had been in sight for more than a day previous.

"Well, what of it?" asked Hermon.

"You are looking upon the Sacred Mountain," was the reply.

CHAPTER XXIII.

BARA-BUDUR.

I HAVE spoken quite often of the wonderful ruins of Java, and must repeat that they are equaled nowhere in the world. They were left by the Hindu conquerors, many long centuries ago. Tjandis as the Javanese call their temples, are quite common in both middle and eastern Java.

The most famous remains on the island were first examined by Eustace and Hermon Hadley, under the guidance of the intelligent Jurak. They lie a little to the west of the right bank of the Praga River, and are known as the temple-ruins of Bara-Budur.

It is hard for me to make clear the surpassing grandeur and magnificence of these ruins. In the first place, a hill one hundred and fifty-four feet high afforded a site for the structure, and the lava blocks which strewed the ground in all directions supplied the material wanted.

Let us consider the dimensions. A square terrace four hundred and ninety seven feet long incloses the hill at a height of fifty feet. Five feet above is another terrace, each side three hundred and sixty-five feet. Eleven feet higher is a third terrace. Then follow four other ramparts and four other terraces, the whole structure being covered by a cupola fifty-two feet in diameter, surrounded by sixteen smaller bell-shaped cupolas.

It is a remarkable peculiarity that on the outside wall of the second enciente are one hundred and four niches, each with its image of Buddha on a lotus throne hewn out of a single block five feet high; and between the niches are sitting figures, man and woman alternately. The inside of the same enciente is adorned with still more astonishing richness. It has nearly six hundred bas-reliefs, representing scenes in the Buddha legend. The space occupied by the bas-reliefs crowded with figures is more than three miles in length. As I have said, the work and skill expended on the great Pyramids of Egypt bear no comparison with that required to complete this sculptured hill-temple of Java.

When our friends reached the ruins of Bara-Budur they found a number of natives moving among them. A person could have spent a week in examining them, but at the end of two hours the party walked toward the Sacred Mountain, which Jurak had pointed out to them some time before.

The mountain itself had nothing remarkable in its appearance, though its elevation is about a mile and a quarter above the level of the sea; but the ruins are so extensive and wonderful that they have been called the Benares of Java, after the holy city of the same name in India.

"Can you tell me why it is called a sacred mountain?" asked Hermon as the three strolled hither and thither, accompanied by Tweak, who frolicked and ran about as though he enjoyed the excursion as much as did the boys themselves. Jurak had inspected them several times, but their splendor impressed even him.

"You know that in India they have holy cities and localities. The ancient conquerors of Java were Hindus, and the custom prevailed among them. The oldest known Javanese inscription speaks of this mountain as holy, and it is still regarded as such by many people."

"The temples are not really on a mountain, as I supposed they were from what you said."

"No. This is the Dieng plateau; but, as I told you, it has long borne the name of a holy or sacred mountain."

As at Bara-Budur, the little party found others beside themselves were present. The boys were surprised and pleased to discover among them a gentleman and his son who were Europeans. Hermon made advances, but the visitors were so churlish that he quickly turned his back—preferring, as he declared, the company of Tweak.

To reach the temples four flights of stone steps

had been made up the mountain from opposite directions, and each flight contains more than a thousand steps. The ruins show that upward of four hundred temples once stood there, and most of them were richly decorated with delicate sculpture-work. For mile after mile the ruins are so numerous that images are found in ditches, in rude walls, and half-imbedded in the mud. The single stairway between Lake Mendjer and Lake Tjebong contains nearly five thousand steps. It is safe to assume that few people can climb that ascent without stopping once or twice to take breath. The entire plateau is drained by an extensive subterranean channel.

Without attempting anything like a full description of the Sacred Mountain and its extraordinary ruins, you will admit that I have said enough to show that they were worth a long pilgrimage to see. When we read or hear one speak of such wonders we are apt to think that if we ever have the chance to visit them we will spend many days in their inspection. We feel as though we can never become weary; but, after all, the sight grows tiresome from its very richness and similarity of detail. You can sit down and enjoy a beautiful painting for a long time, but if you start to examine a gallery containing hundreds and thousands, the finest of them will soon cease to interest you.

So it was in the case of Eustace and Hermon. They strolled about, expressing their wonder at almost every step, and frequently pausing to view more closely the marvelous exhibition, until, when the afternoon was pretty well along, they began to tire. The sun was sinking in the sky, and they sat down on some of the bowlders to discuss what was best to do.

They had wandered so far to the westward that no other persons were near them. They could see people in the distance, like dark specks moving about, but the friends were alone and safe from interruption. While the boys rested and Tweak darted hither and thither, Jurak stood erect, leaning on his gun and gazing in the direction of the main collection of ruined temples. In the declining rays of the sun they were gilded with impressive splendor, and the dreamy, far-away expression in the eyes of the native showed he was sunk in deep thought.

More than likely the religious side of his nature was stirred by the sight and the recollections it called up; and as he viewed the handiwork of men who had perished from earth centuries before, and reflected that this was done as an expression of worship of Him who rules over all, Jurak would have been stolid indeed had not his feelings been stirred and his heart touched.

CHAPTER XXIV.

A WELL-VENTILATED APARTMENT.

"WELL," said Eustace, addressing his cousin at his side, "you have crossed the island of Java, as may be said, and stand on the Sacred Mountain. Do you feel repaid?"

"Aye, ten times over. I shall recall this jaunt through the country with delight, no matter where I am nor how many years from now it may be. Pretty soon we must go back to school and settle down to hard work, so we will enjoy our vacation while we may."

"Shall we return afoot or on horseback?"

"I would prefer to go on foot and to double on the route we followed in coming here."

"I know it will be pleasant to pass through the villages where you are known as the mighty hunter that slew the tiger, and where they never grow tired of admiring you."

Hermon laughed and shook his head.

"You know it isn't that; but they promised me the leopard-skin, and I would like to buy that of the tiger from Jurak."

The latter, hearing his name pronounced, looked

down at the boys and asked what had been said. When Hermon repeated his words he replied:

"It will be much pleasure to me to present you with it. You slew the terrible animal and it belongs to you."

"It might if I had removed the skin from him; but the work is worth something, and I will take it if you will accept pay for it."

Jurak shook his head.

"You cannot pay me for it; but," the cunning fellow added, "Myeta would not be grieved if some little token was given her, that she might be helped to remember your visit."

Eustace winked at his cousin, who nodded slightly and said:

"Myeta was kind to us, and I would always be sorry if she would not allow me to present her with something; so we will consider the matter settled. I don't want to be looked upon as the Eighth Wonder, for any one with common sense cannot fail to see that there is nothing so remarkable in what I did."

"There may not be as much as some are inclined to claim for you; but after all it *was* remarkable, say what you please."

"The good luck was; but a tiger stops in front of me in the best position possible he could take to be killed; I aim, pull the trigger, and he falls. Then the leopard is within a few feet when he jumps from a limb, and with the same excellent gun I wound him so that he soon dies. You would have done just as well if the same chances had come to you."

"I think so," remarked Jurak, with a nod of his head.

"Perhaps I might," said Eustace; "but the chances didn't come to me, so what's the use of talking of what might have happened? I didn't do it and you did. You deserve credit for your coolness, for many an old hunter cannot control his nerves when he finds a tiger as close as that one was to you."

"Maybe if I had known as much about him as an old hunter I also would have trembled."

"But when you go to school with me and show the leopard and tiger skins, you will be warranted in feeling proud as you inform your friends that you killed the beasts without help."

"Of course I shall; but there will be one drawback."

"What's that?"

"None of them will believe me!"

"But I will assure them that I know what you say is a fact."

"And how will *that* help matters?" asked Hermon.

Eustace laughed over the tap he received and replied:

"I shall be careful where I relate the story; it will be among the people with whom I am well acquainted."

"And the danger will be all the greater, because of necessity they must know *you*. However, that is a small matter. I suppose I could get the skins without going for them, for Jurak could obtain them; but the jaunt is pleasant enough to walk, and I think it will do us more good than if we ride."

"I have no choice," said Eustace, "and as there is plenty of time we won't hurry. Jurak can show us some new places and scenes, and you will have the more wonderful stories to tell when you get home."

"All of which being so, I beg to remark that it is growing dark, and it is time we hunted up a hotel for the night."

As the party meant to start back on the morrow, it struck the cousins that the proper thing to do was to pass the night among the ruins. There was nothing to be feared from any change in the weather, and their blankets were all-sufficient to protect them against the coolness which was felt at that elevation. Beside, the ruins themselves afforded all the shelter they could want. It was agreed, therefore, that they would stay in the immediate vicinity.

The only question that caused any thought was as to food. All three were quite hungry and there was none near them; but Jurak told the boys to locate the camp and he would bring them their supper.

With this remark their guide, as he may be called, walked away, and the boys set about finding a suitable spot. As a matter of course there was little or no difficulty experienced. They stopped near the ruins of what had once been an annex to a much larger building, though it was impossible to guess the real purpose for which it was originally intended. It was made of volcanic stone, and about one-fourth had lain buried for ages under the ashes thrown out by some stupendous volcanic eruption. The boys walked through an archway a couple of yards in width, but from the cause mentioned they had to stoop as they passed under the arch. It was so dark inside that Eustace struck several wax matches and held them above his head. By their flickering light they saw they were within a space nearly twenty feet square, on each side of which were three archways similar to the one through which they had entered.

"There's one advantage," remarked Hermon—"we shall have the best kind of ventilation."

"Yes, I don't see that there is any difference between this and all outdoors. It is so dark and gloomy in here that we must have a fire, so let's gather the wood."

It required a short time only for the boys to gather all the fuel they would be likely to need. It was so dark within the temple that they made haste to start the fire, which speedily illuminated the interior with its ruddy glow. It was then noticed that the ceiling was of plain, smooth stone, and about a dozen feet above their heads. As for the rest, their eyes discovered nothing of special interest. They spread out their blankets and sat down to await the coming of Jurak with their supper.

"It seems to me," said Hermon, looking around on the interior, "that this is a good place for serpents."

"It's a good place for anything that takes a fancy to it; therefore it's a good place for *us*."

"But I have been told that in India and other tropical countries where poisonous snakes abound they are fond of hiding in ruins and old buildings."

"Of course they are, for they are so plenty in that delightful country that you find them everywhere. When twenty thousand people die every year from snake bites, the poor reptiles must crowd into any place they find."

"There are a number of venomous serpents belonging to Java," said Hermon, "but from my experience they are not numerous."

"No; I have tramped for days through the woods without seeing one. I don't believe there are any near us."

Hermon was naturally relieved to hear his cousin speak with such positiveness, and when Jurak came in shortly afterward with an abundance and variety of fruit, and all began to eat, they were in high spirits.

Something more substantial would have been relished by the boys, while the native missed his rice, which the Javan holds in higher regard than any other article of food; but it was no hardship to make their evening meal from such a supply, and when Jurak lit his pipe and the smoke curled slowly upward he was the picture of lazy enjoyment.

But, as before, he became abstracted and thoughtful. He took no part in the conversation, and was either gazing at the glowing embers at his feet or looking absently at the stony ceiling, which reflected the glow of the fire.

"Jurak," called out Hermon from the other side of the blaze, "Eustace and I have noticed that you have been absent-minded and thoughtful all day. What's the matter?"

The native looked around with a half-startled look as though he had been doing something amiss, and then smiled in a way which proved that he was in anything but a mirthful mood. He was silent a minute or two longer and then said:

"I knew I would feel sad when I came to the Sacred Mountain."

"And why so?" asked Eustace. "You have been cheerful at times, and chatted like yourself when we were eating supper; but both of us see that something is on your mind."

"It looks as if unpleasant associations are connected with the place," suggested Hermon.

"You have spoken the truth. I have been here before; the memory is sad."

Looking from the face of one boy to another, he noticed their expectant expressions.

"Ten years ago I came to the Sacred Mountain and the Bara-Budur. I rode on horseback, and was with a missionary and his little son and daughter. The boy was fourteen, the girl fifteen. They were lovely children, and each rode the pony as well as I could do. The missionary was a good man, and he felt much pleasure in looking over everything that could be seen. He wrote much in a little book he carried and asked many questions, and I was pleased to answer him. When it came night he wanted to go down to the village to stay; but his son and daughter thought it would be nicer to camp among the ruined temples, as we are doing. He was so kind that he hardly ever refused them anything, and so he agreed to do so. I took care of the horses and brought them fruit to eat, just as I did for you, and we built our fire among the ruins."

"Did they camp in here?" asked Eustace.

"I thought at first it was the place," replied Jurak, again casting his eyes over the ceiling and surroundings, "but it is not. If it had been, I would not have stayed here."

"Why not?"

"Listen, and I will tell you. They had more luggage with them than have you, and they made much preparation when they went into camp. The brother and sister were tired, and they wrapped themselves up for the night and were soon asleep. The missionary sat a long time talking with me and writing in his small book. He was doing that when I put out my pipe and also lay down to sleep. He told me afterward that he sat up a half hour longer and then he, too, lay down."

"Then you were all asleep and had no guard?"

"Yes; it was not needed that some one should keep watch."

"How did you do when in the woods?"

"They never stayed over night in the woods, as we have done. The brother and sister often wanted to do so; but the father would not consent and we slept in the houses. There was nothing to fear here, and the missionary told me that he had no thought of danger when he lay down."

It was clear from the manner of Jurak that he had a singular occurrence to relate, and the boys closely listened.

"I was the first one to awake in the morning, and I began stirring about and bringing water and getting the fire ready for our morning meal. While I was doing so the father and daughter aroused themselves, and then it was noticed that the boy was absent.

"We thought nothing of that until an hour had gone by without any one seeing him. The father went outside the ruins and called, but no answer came to him; and then we all were alarmed, and set out to find the boy.

"Well," added Jurak with a sigh, "we hunted and inquired, and engaged other persons to hunt everywhere, but he never was found. Not the least trace of him ever came to light. He was a bright, good boy, and when I think of him I cannot help feeling sorrowful."

CHAPTER XXV.

MISSING.

NO WONDER that Jurak felt grieved when he recalled the affecting story.

"At last," he added, "the father and sister were obliged to go back to Batavia without gaining tidings of the missing boy. They were broken-hearted."

"I do not wonder that they were. What became of the missionary?"

"He came to me two years afterward, and I went to the Sacred Mountain to search with him for his boy. He had had other men looking ever since we were there, and offered a large reward; but it was no use. The father looked twenty years older than when I saw him before. We did all we could, but the end was the same. He went back more broken-hearted than ever, and I have never seen him since."

"That is a sad story, indeed," said Hermon. "Its remembrance must bring you grief. But what could have become of the boy?"

"Do you think, if I could answer that question, I would not have done so when the father and the sister asked me so many times?"

"Of course not; but there must have been some theories which occurred to you all. Such things don't take place without a cause that ought to be known to some one."

"The father thought his son had wandered off to the woods in his sleep and had been devoured by some wild animal. Sometimes I think the father was right. Another old acquaintance, Marriavo (I got our supper at his hut), had the oddest explanation of all. You know we are not very far from the ocean. He said he had seen some rough-looking sailor-men in the neighborhood for several days, and he thought they had stolen the boy."

"Is it not possible he was right?"

Jurak shook his head.

"It could not be. Would any party of men come into a place and steal a boy and not disturb his beautiful sister who lay asleep, with her father, near at hand? Why would they have taken the boy and left the others and the property? Why would they have stolen the boy, anyway? No; it is not possible; the boy did not disappear *that* way."

"How, then, was it?"

"I cannot tell. I have thought of a good many ways, and it seems most likely that it was as the father believed—he wandered off in the darkness and was slain by some wild animal."

Eustace knew of the journey which Jurak made with the missionary ten years before, but he had never heard anything of the strange disappearance of the boy. Like his cousin, he was much impressed by the narration, and it was natural that they should speculate a good deal in their hunt for some rational explanation of the occurrence.

Jurak listened but made no comment. He refilled his pipe, and fixing his eyes on the fire, which was beginning to smoulder, slowly puffed, and sank again into a dreamy abstraction. The boys felt that he had exerted himself to tell them of the occurrence, and they did not question him any further.

It was quite late before the cousins felt drowsy and made their preparations to sleep.

"I think that boy must have been a somnambulist," remarked Hermon, after he and Eustace had wrapped their blankets about them and were waiting for the coming of slumber; "he strayed off, just as the father believed."

Eustace assented, and after some further talk the two fell into slumber.

Jurak continued smoking a long time. He made no move to replenish the fire, which burned lower and lower, until he could hardly see the forms of the unconscious lads stretched out on the floor of the old ruin. On one side a few of the moon's rays penetrated the place, but the orb was so high in the heavens that only a few inches of the interior were illuminated.

At last the native knocked the ashes from his pipe, and with a sigh slowly rose to his feet and walked outside, where he stood a long time in contemplation.

The scene was impressive, the bright moon lighting up the ruined temples and buildings like a city whose inhabitants were wrapped in slumber. It was hard for one to realize, while gazing around him, that the men who had builded these wonderful structures had been dust for so many centuries—that they had perished so utterly from the earth that the piles of stone were the only evidences they ever existed. No living man or animal was seen, though people dwelt at no great distance in their primitive structures.

A faint, almost inaudible murmur was borne to Jurak on the gentle breeze which swept over the island. It was the voice of the Indian Ocean, which miles away beat against the rocky shores of Java, as it had done ever since the island was torn from the mainland of the continent by one of the mightiest upheavals of the world.

Finally, as if wearied with mental oppression, the native turned about, and wrapping his thick blanket around him, lay down on the hard floor close to the embers, which had almost died out. It was a long time before he slept; but when he did so, his slumber was so deep that he did not awaken until daylight lit up the place.

Even then he was aroused by Eustace, who shook him vigorously by the shoulder.

"Jurak," said he, "you sleep sound. I have been awake fully a half hour. *Hermon is gone!*"

"What!" exclaimed the guide, springing to his feet. "Where is he?"

"I cannot tell. I am greatly alarmed. Can it be that any ill has befallen him?"

"We will soon find out," replied Jurak, unable to hide his emotions. "I am afraid something *is* wrong."

Under ordinary circumstances the absence of Hermon Hadley would have caused no alarm until it had lasted a long time, but it will be understood why both Eustace and Jurak were in great distress from the first.

In every respect the disappearance of the youth corresponded, so far, to that of the missionary's son, of whom no trace was ever found. Jurak and Eustace did not wish to believe it was due to the same cause, but how could they help thinking so?

Eustace related that when he slept his brain ran riot over the strange story of the other boy. At the moment he opened his eyes he felt that something was amiss. His face was turned toward Jurak, and raising his head, he noticed that he was sleeping so heavily that his breathing was plainly heard. Then he turned like a flash toward the spot where his cousin had lain down near him. There was his blanket, but Hermon, with his gun and everything else, was gone.

Forcing down the dreadful fear that shot through him, Eustace did not wait to awaken Jurak, but dashed out of the ruins and ran hither and thither, signaling to his cousin by means of the sharp whistling they often used when separated from each other. Of course there were no answers, and then he hurried back and aroused the guide.

The two almost instantly parted company and pushed their search for fully a quarter of a mile in every direction. Eustace, as you remember, could speak the Javanese as well as the native, and he asked questions of every one whom he met. To the west, north and east stretched the ruins and remnants of buildings, through which the youth hunted with the utmost vigor. To the south, a quarter of a mile distant, was a native village.

While hurrying among the old temples, Eustace saw people here and there. He approached all within reach and asked whether they had seen anything of Hermon. The natives, some of whom seemed to make their home among the ruins, listened closely, and would have been glad to help the youth, whose distress was so apparent; but in every case they gravely shook their heads and replied that they had seen nothing of the missing one. Pausing only long enough to add that any one who found him would be liberally paid, Eustace hurried on to question the next one whom he saw.

Every minute or two he would stop, and doubling two of his fingers and placing them in his mouth, emitted a whistle which sounded like that of a steam-engine. Then he paused for the reply; but none came.

While he was thus employed Jurak gave most of his attention to the native village, just as he hunted ten years before in the same place for the lost son of the missionary. The house first visited was that of Marriavo, where he got the food the night before, and whose owner, long ago, was so sure the lost lad had been stolen by a number of piratical sailors who were skulking about the village.

He listened with close attention to the story told, and then, shaking his head, made the amazing statement:

"Alack! he, too, was taken by the pirates!"

"Why do you think so?" asked Jurak.

"I have seen the same men in our village. One of them stopped at my door yesterday, and I gave him rice and tea."

"But why should they steal one boy and leave the other?"

"If I could answer questions I would do so. I am sure they ran away with the lad ten years ago; and now they have taken another to fill his place, for the other is a man or he is dead."

Finding his friend had no real knowledge, Jurak came out of his hut and pressed his inquiries elsewhere. He failed to get the slightest knowledge; no one had seen or heard of the boy.

As the native walked thoughtfully back to the ruined temple, where they had spent the previous night, he asked himself whether it was possible Marriavo was right in accounting for the absence of Hermon Hadley. He was no more positive ten years previous than he was now, and in each instance he spoke of the rough, sinister-looking sailors whom he had seen loitering through the village and the vicinity. Could it be they had carried Hermon away?

Without pausing to consider fully the improbability of such being the case Jurak retraced his steps to the village and made inquiries about the sailors. He not only learned that no one else had seen any such person (proof, under the circumstances, that none had been present), but Marriavo was looked upon by all as not exactly right in his head. The sailors which he fancied he saw had no existence.

It was almost noon when Jurak, having pressed his search in every place possible, went back to the ruins where Eustace sat on a bowlder awaiting him. The disconsolate looks of the youth told the story without any questioning. Heaving a deep sigh, Eustace rose to his feet.

"Jurak, doesn't this beat everything of which you ever heard?"

"This and the other case does," was the reply of the native, who dropped the butt of his gun on the ground, and folding his arms on the muzzle, took an attitude of deep thought.

"What do you think has become of him?"

"I know no more about it than the baby Woo Woo at the home of Dati, where we stayed over night."

"You said the other boy might have been carried off by a tiger or wild beast. I have searched all around this place, and can find no traces of the tracks of any animal."

"I never thought the wild beasts came into the ruins and stole the boy; *that* could not have taken place without arousing the rest of us; but I believed the lad had strayed away in his sleep and had then been seized by a savage animal."

"Do you think it was so with Hermon?"

"There are many reasons, which you must know, why it isn't likely that it was; but can you give any other cause for his absence that has not a great many more likely reasons why it is not true?"

CHAPTER XXVI.

CALLING UPON THE CANINES.

BOTH Jurak and Eustace had scrutinized with the closest attention the earth around the old structure in which they spent the night. They bent over until their faces were as low as their knees, and they looked like a couple of Shawanoes hunting for the trail of some enemy. They were unable to detect the slightest imprint; but had their vision been trained to the fineness of that of the American Indian they would have made an important discovery.

"There's one thing certain," exclaimed Eustace firmly, closing his lips as he straightened up; "I'll never go back home without Hermon."

This declaration was characteristic of him who made it; but it may well be said that it did nothing toward finding the missing boy, nor did it help explain why he was not found.

"I will stay with you till we learn the truth."

But while the words were in the mouth of the native he added that the missionary and his daughter were just as positive when the other boy was lost. The sister declared with sobs that she would never leave the Sacred Mountain without him. That was ten years ago, and she could not keep her word.

But Eustace shook his head.

"That was said by a *girl*. Do you suppose I will ever walk into our house, and when Hermon's mother asks me where he is will tell her I don't know?"

"You may have to do so."

"No, sir," said Eustace, more decidedly than ever. The idea that he can slip away from the camp-fire like that and the whole Javan population can't find him is absurd. There's one thing that makes me mad," added the youth.

"What is that?"

"I brought along that miserable pup Tweak, hoping he would prove of some use. Hermon laughed at him from the first, but I stood up for the creature. He was on hand yesterday to get something to eat, but then he went away, and nothing has been seen of him since. If he had been here in camp last night with us he would have been of some help. If there had been any violence he would have given warning, or if Hermon had started off while asleep he would have followed him, or barked and awakened the rest of us."

"I don't think the dog knew enough."

"The dullest dog in the world would have been equal to that. But Tweak has not shown up to-day, and I don't suppose anything more will be seen of him till he wants something to eat."

"What good could he do if he was here now?"

"I have been wondering whether he could not take the trail, as they call it in America, of Hermon, and follow it. They have dogs in England and America called bloodhounds which would need only a sniff at the blanket there, and they would follow Hermon everywhere. No matter how many trails crossed his—there might be hundreds—the bloodhound could not be shaken off unless the person took to water. He would follow Hermon through woods, over rocks, on the highway—anywhere and everywhere he went."

Eustace, as you know, spoke the truth; but Jurak did not believe a word of it. He merely remarked:

"Tweak isn't that kind of a dog, and I have never seen any such."

"I saw a couple belonging to an English gentleman in Batavia, but I hoped there were some dogs in the village that knew enough to take the trail of Hermon from this point."

"I do not think so, but perhaps there may be. We can do nothing better than try."

With praiseworthy promptness the two acted on the idea. There are plenty of dogs in Java, though the breeds are not superior. Our friends made their way at once to the village and procured three of the best canines that could be found.

By this time, too, it had become generally known that the youth had vanished in an unaccountable manner during the night, leaving nothing to show what had become of him. Jurak had taken pains to tell that whoever found the lad would be liberally rewarded. As a consequence, when they returned with the dogs they were accompanied by a half-dozen young men and youths who meant to take the cue from the action of the canines, and then press the search with all the skill and vigor at their command.

Three times as many were already pushing their investigations in other quarters, some of them being far away in the forest, while others were searching among the ruins as closely as if on the hunt for some jewel of great value.

The action of the dogs raised the hopes of Hermon. They sniffed the ground and ran back and forth, as if they understood what was expected of them. One uttered a sharp bark, threw his nose in the air, and made a break as if he had uncovered the game. The others imitated his actions, and the natives, including Jurak and Eustace, dashed after them all.

The dogs galloped directly eastward, leaping over stones and bowlders, dashing through undergrowth, and running around such obstacles as interposed across their path. Once, when Eustace stepped on something, he looked down and saw that he had planted his foot directly on the nose of a sculptured image which was lying on its back and staring toward the sky.

The dogs hastened until, when only a short distance away, they came upon a well-beaten path. Turning to the right they trotted briskly forward, emitting a yelp or two, and the men in a long string dashed after them. But all at once the leading mongrel, with another sharp cry, wheeled about, the other two doing the same, and charged backward among the legs of the men directly behind them. Several of the natives became so entangled that they stumbled and fell, whereat there was much more barking and shouting, and for a few minutes everything was in confusion.

The discouraging feature of the proceeding was its proof that if the dogs had been following the trail of Hermon Hadley they had irrecoverably lost it.

CHAPTER XXVII.

IN THE NIGHT-TIME.

THE END of it all was that when the middle of the afternoon came the fate of Hermon Hadley was as much a mystery as ever, and the dogs that had been employed to take the scent had shown they were unable to do so. To one so alarmed and impatient as Eustace was, it seemed that a cloud of stupidity had fallen on men and beasts alike. Here were people who had lived all their lives among the ruins, as may be said, and yet they seemed to know nothing about them. The dogs which ought to have been able to follow the trail were equally stupid, and Eustace whispered to himself, with a shudder, that it looked as if the strange story told by Jurak was to be repeated.

"Suppose the days and nights go by without bringing me the first clew to his whereabouts, and I have to go back at last without him—what shall I say to his mother? When she hears that we lost Hermon, and don't know what has become of him, will she be content to stay at home and never make search? No; she will hasten to this place, and never give up till she learns everything."

The result of all these reflections was a stronger determination than ever to learn the truth.

It was exasperating to reflect that it seemed impossible to do anything toward solving the mystery. It was like trying to climb an insurmountable wall. It was annoying to look at Jurak and reflect that, old and experienced as he was, he was as baffled as if he were a child.

Eustace had not eaten a mouthful since the night before, and did not wish anything. His feverish impatience to learn what had become of his beloved friend rendered him insensible for the time to fatigue and hunger, and would not allow him to stay quiet more than a few minutes in any place. Believing that more could be done by separating, he kept far away from Jurak, while the natives were left to prosecute the search as they preferred.

It was agreed by Jurak and Eustace that if either found any clew he would signal to the other by firing his gun. As the night approached it was arranged also that a fire should be kindled on the outside of the ruins where they spent the previous night, and the place was to be considered the headquarters of the searchers; that is, in case any one of the natives learned anything he would instantly return to the camp, from which he would signal to the others by means of the cries peculiar to the people when hunting.

When darkness closed in over the forest, village

and ruins, Eustace Hadley for the first time began to feel something like despair. He had been feverish and restless during the day, and had tramped so much that now he became tired, and was faint from the want of food. Jurak compelled him to eat quite heartily and to drink some coffee which, rather curiously, was prepared by Marriavo, who brought it from his home and warmed it over the flames that were burning outside the ruins.

He had refused to join in the hunt during the day, for he declared it was useless. The boy had not wandered off in the woods nor been slain by wild beasts; he had been picked up bodily by the evil-looking sailors, carried to the sea-shore, and then put aboard ship and taken off.

Eustace more than once found himself debating the question whether it was not possible the old fellow was right, after all. Despite the improbability of such being the case he had selected two of the swiftest runners, and directed them to make all haste by the nearest highway to the sea-shore. They were to push their inquiries without cessation, and in case anything at all was learned were to apply to the authorities without delay. They would have little trouble in securing all the help they needed, for, as I have told you, Java is one of the best governed countries in the world.

Sober reflection convinced the youth that it was a fool's errand on which the natives were sent, but had it not been done he would have reproached himself ever afterward for not doing everything that promised success even in the slightest degree. He told Jurak nothing of the plan named, for he knew well enough that the sagacious native had no faith in it.

Hitherto the search had been prosecuted without any regularity or system, for the good reason that it was out of the power of any one to do so. Had the missing boy been known to be in the water or lost in the woods, the parties could have divided and pressed the hunt according to some plan; but, as it was, no person could know where to make search.

It being the dry season, or summer-time, the weather continued of the same pleasant, equable character; and at quite a late hour the moon, unobscured by clouds, rose in the heavens. Most of the natives had gone to their homes, promising to resume the search on the morrow. Naturally, they could see no reason to expect success while darkness lasted. Two of them stayed by the camp-fire and kept it going, while Jurak and Eustace wandered here and there, with no particular destination in view. The youth had been on his feet nearly the entire day, and as has been said, when darkness closed in he was worn out. He had sat down by the fire intending to sleep, as most of the natives were doing; but his anxiety was too great, and he finally sprang to his feet, gun in hand, and began picking his way toward the more mountainous region which lay to the north.

His energy lasted half an hour or more, when he sat down on a fallen tree and asked himself what was best to do—if, indeed, he could do anything. The trees were growing thickly around him, so that his view was not very extended in any direction. On the right hand, a few rods off, could be distinguished a pile of stones, some of them so shapely that no doubt they were parts of images that had once graced the temples whose ruins covered scores of square miles of surface.

"Where can poor Hermon be?" he asked himself for the hundredth time. "Last night we were sitting around the camp-fire listening to Jurak's strange story, never dreaming that anything of the kind could happen to either of us. I wonder how it is that he was taken and I left? Of course if they wanted the smartest and best-looking they would select him; but it is curious that neither Jurak nor I was molested."

Eustace sat on the fallen tree for some fifteen minutes longer, a prey to the most distressing emotions. He kept up his courage remarkably well during most of the day, but the time had come when he could not drive away the despair that was slowly filling his heart.

Thus he sat when the clear report of a rifle struck his ear. It came from the direction of "headquartors," and there could be no doubt that it was fired by Jurak as a signal that he had gained some tidings of the missing Hermon.

The youth bounded from the log as though he had heard the hiss of a serpent behind it. He could not see the ruined temple where the fire was burning, but he knew the direction, and it was not far away.

The wonder was that he did not break his neck in his struggles to pass the intervening distance. He frequently fell on his hands and knees, and his garments suffered much; but he did not mind such slight troubles, and in much less time than would be supposed he reached his destination.

The two natives were throwing fuel on the fire, while Jurak stood erect, gun in hand, looking eagerly around in the gloom, as though wondering why Eustace did not appear. Before he saw the boy the latter called out:

"Halloo, Jurak! Is it good or bad news?"

"I cannot say."

"Then it is not bad."

"I hope not; but I cannot be sure."

"What is it?"

By way of answer the native held a gun aloft, so that the light from the fire fell on it. One look was enough to tell Eustace that it was Hermon's gun.

"Where did you get it?" eagerly asked the youth.

"I found it only a little ways to the west, where

you and I and the others must have walked by it many times to-day."

"How long ago was that?"

"I cannot say—but it was a little while."

"Did you fire your gun?"

"Not until I reached the camp-fire; I wanted to look over the weapon by the light of the blaze. I expected to find you here, and when I saw you not I fired my gun."

Eustace reached out and took the weapon. It was with strange feelings that he grasped the stock and barrel which he had held so often, and which belonged to his beloved cousin.

"Halloo, it's broken!" he abruptly called out.

Jurak stepped forward, and noticed for the first time that the hammer had been snapped off as though struck a blow by some heavy object. The gun was useless.

This was the only injury the piece had suffered, and a close inspection could detect nothing else worth noting.

"Where did you find it, Jurak?"

The native turned about and pointed off in the gloom.

"Over yonder. It was covered with dirt and leaves, so that it would not have been seen had I not struck it with my foot. Even then it was not recognized in the darkness until I stooped over. I would not have taken the pains to do even that had

I not caught the gleam of the silver on the stock, on which a little moonlight fell. I was sure it was the gun of Hermon, and brought it back to the fire."

"Did you make no search around there?"

"I did not forget that. I called his name and moved softly in many directions. I was disappointed, for that has been the rule all day."

The recovery of the rifle naturally set the two to speculating and guessing again. They sought to explain it in many ways; but as none of the theories were right, it is not necessary to mention them here. They were compelled to admit that nothing could be done that night, and weary and saddened, Eustace wrapped his blanket about his shoulders and lie down, leaving Jurak seated and smoking his pipe, while the two natives were squatted on the ground a short distance off, talking together with as much animation as though they were arranging to go on some pleasure-excursion.

It will be admitted by all that the outlook was bad. The chances were very much against Hermon Hadley. He had now been gone twenty-four hours. If alive, it would seem impossible that he could have remained undiscovered so long, since he would be sure to put forth every possible effort to rejoin or at least to communicate with his friends.

Eustace was as determined as ever to press the search on the morrow, but he could not fail to see that the ground for hope was of the most fragile nature.

He was gradually sinking into unconsciousness when he caught the faint footfall of some animal. Instantly his senses were on the alert and he raised his head. The couple were still sitting on the ground, talking in low tones; but Jurak had also lain down, and seemed to be asleep.

At the same moment he observed Tweak, the dog that had been gone so long, standing near, wagging his tail and looking wistfully at him.

"Why didn't you come back long ago?" asked Eustace angrily. "If you had been here last night you might have saved him; but now all you want is something to eat, which you haven't earned and sha'n't have."

The youth dropped his head on his arm, with his back to the canine. The latter came closer and whined; then he walked around the prostrate boy and continued his whining. Eustace raised up again and asked:

"What's the matter, Tweak?"

The dog trotted a few steps, then whined and ran back again.

"He wants me to follow him," said the lad in an awed whisper, as he rose to a sitting position. "Can it be he has found out something about Hermon? I shall soon know!" he exclaimed, rising hastily to his feet and starting off under the guidance of Tweak.

CHAPTER XXVIII.

HOW IT CAME ABOUT.

NOW THAT the friends of Hermon Hadley have hunted so long for him without success, let us take up the search and see what we can do.

It is more than likely that every one of my readers has learned by experience how apt one is to dream of that which has unusually interested him during the day or evening before going to sleep. Have you not puzzled your brain over some problem until in sheer weariness you gave it up and went to bed? And has not the brain continued its work after your eyes were closed, and perhaps straightened out the whole thing for you before daylight?

If this has not happened, you have dreamed vividly over some matter which absorbed your thoughts when last awake, and perhaps you have cried out and struggled in your sleep. The affecting incident told by Jurak produced a strong impression on Hermon. He lay awake a long time, and when his eyes closed his brain was topsy-turvy. As near as can be learned it was near morning when he suddenly rose from his blanket, and rifle in hand walked so softly out of the ruins where they were lying that neither Jurak nor Eustace was disturbed.

Hermon was not a somnambulist, and had never been known to walk in his sleep; but he did so when he stole out of the ruins and in the dim moonlight began picking his way toward the mountainous region which lay to the north. He was doing so in obedience to a strange impulse which, in his peculiar state, he could not refuse to obey.

He had not gone far when the conviction came over him that he was doing wrong in taking his gun with him. He laid it on the ground, therefore, and walked forward with the care and skill which persons in his condition often show. All at once, and without the least warning, he dropped through an opening in the ground and disappeared.

This opening was just large enough to admit his body, and was so entirely concealed by the bushes growing near that a person could step within a few inches of its margin (as had been the case) without suspecting its existence. Hermon fell about fifty feet, and it need not be said that he was thoroughly awakened when he landed.

Had not his fall been broken he must have been killed, for the distance was enough to destroy life in any one; but he struck against the side of the chamber in going down, loosening so much dirt, vines and debris that it was driven in front of him, and formed a cushion on which he was sitting with outspread feet when his descent suddenly stopped.

Who can imagine the feelings of the youth when his senses came to him? He had been scratched and bruised and jarred; his hat had fallen off, and he had gone down into some place of whose nature he had not the remotest idea. He would have believed he was still dreaming but for the reminders in his body of many pains.

"What can this mean?" he asked himself, sitting still and peering in every direction. But the darkness was without the slightest ray, and when he looked upward he could see nothing of the opening through which he must have entered. It was so shaded by grass and bushes that none of the faint moonlight could reach it.

But Eustace and Hermon always carried matches with them, and the younger had a number of waxen ones in a small rubber case. Drawing this forth, he struck one on the bottom, and as it blazed out held it above his head and again looked about him.

The radius of illumination was not large, but it told an interesting story. He was in a chamber or under-ground apartment some fifty feet wide, and as he afterward learned (for the single match could not reveal it) more than twice as long. It was quite regular in its outlines, and in many places he could trace the forms of the rough stones which composed the walls. On one side the wall had crumbled a great deal, so that it lost much of its perpendicularity. It had also been greatly overrun with clinging vines. Fortunately the lad fell among these, and, as has been shown, they broke his descent. Many of the vines had been torn loose and lay at the bottom of the cavern, or whatever it may be called.

By the time Hermon learned this much his match went out. He struck another, and then, knowing there were no more than a dozen in the box, saw he must be sparing in their use. It would be a great calamity to be left in utter darkness.

The chamber must have been constructed ages before by the Hindu occupants of the country, and of course bore some relation to the temples above ground. No doubt many other similar excavations existed in the vicinity. The vines which grew over the walls on every side had been there so long that some of the trunks were several inches in diameter, and all were clustered with dead twigs and branches which would burn readily.

With the help of another match he gathered a lot of this stuff together, and holding the flame to its base for the few minutes it burned, succeeded in igniting it. As the flame spread he collected more, until by working actively he soon had a large pile. One disadvantage was that it burned out very rapidly, and large as was the amount gathered, it could not last a long time. So far as he could see there was nothing else within reach that could serve as fuel.

But the blaze fully lit up the interior, and for the first time he gained a fair view of it. Stepping away from the blaze, so that the glare did not interfere, his eyes roamed over the interior.

While the sides of the chamber had been fashioned with some regard to proportion, the roof was very irregular. It was broken and jagged, as if made by nature alone. No doubt advantage had been taken of the work of some earthquake by those who had given the place its shape.

It took some time before the youth could locate the opening through which he had entered, but finally he discovered it far up in the roof. It was smaller than he supposed, but, as he grimly said to himself, it was large enough to answer.

Looking down at his clothing, Hermon saw they were sadly torn and soiled, while his hands and face were scratched, and the aches and pains in his limbs proved how roughly he had been handled. He was bright-witted enough to suspect the truth.

"I have been dreaming by the fire, and got up while all of us were asleep and wandered over the mountain until I fell into this place. I was never known to rise in my sleep before, but it must have been because I felt so bad over that story Jurak told."

Naturally his absorbing thought was as to how he was to get out of the place.

"Neither Jurak nor Eustace know of my going

away, for if they had they wouldn't have let me go. Since those vines helped me down, I wonder whether they won't help me up?"

"Stepping to the wall, he found they were thick, and some of them clung as tightly to the stones as if wired fast. No doubt they would support much weight, though if they were torn loose at the top, as many of them had been by the violence with which he struck them, they could be readily pulled away.

But with his native pluck and energy the young American began at once to climb the crumbling wall toward the narrow opening, fifty feet above. To one accustomed to such activity and trials of strength this was not difficult, and he went upward almost as if he was climbing the side of a house by means of a ladder.

This looked hopeful; but the misgivings with which he started increased with every foot he ascended until he had climbed a little more than half way. The higher he went, the sparser and weaker became the vines. Though, as is the case with under-ground vegetation, they inclined toward the opening through which a little daylight sometimes made its way, and though this growth had continued from time immemorial, it was not vigorous enough near the roof to support the weight of the climber. A good deal of it had been torn away by his forcible entrance, and that which was left was much weakened.

Harmon Hadley lost in the cave.—(See page 235.)

Hermon went upward with the utmost care, testing the rib of each creeping vine before trusting a part of his weight upon it. The greatest caution was needed in the use of hands and feet, as any one can testify who has undertaken anything like such a task.

But half-way up he took warning from the tearing away of a vine which he seized. Holding himself stationary, he looked to the roof. which was plainly shown by the reflection of the fire burning below.

One glance was enough to prove the utter impossibility of the task he had undertaken. Before he could reach a point within a dozen feet of the top he would be without any support at all. If the vines were twice as vigorous it would be beyond his power to make his way through the opening if he should reach it, since there would necessarily be a moment when he would have to let go with his hands, raise himself upward on his feet, and then catch the sides of the opening and draw himself through.

In some places, as I have shown, the side of the wall had crumbled so much that it lacked coherence, so that in one sense it may be said the vines themselves supported the wall. Should Hermon climb further, more than likely a sudden yielding would follow, and he would be less fortunate in his fall than before.

"It's no use," he said, with less discouragement than would be supposed; "I can't get out of this place by *that* door."

And then, with the same care he had used from the first, he began his descent. This became easier the lower he went, until in a few minutes he leaped lightly to the ground and threw more fuel on the flames.

Though the youth was in a most trying situation he did not despair. Because he saw no ready way out of his difficulty he was not prepared to say that none such existed. He believed that while there was life there was hope.

"I wonder whether there is any other egress than that by which I entered?" he mused, staring at the gloomy walls in the hope of finding something of the kind.

Throwing on enough more fuel to send an illumination the entire length of the interior, he moved rapidly hither and thither in his search. He thought possibly there was a series of under-ground chambers connected with each other, though even if he should be able to pass from the one into another it was not likely his situation would be improved thereby.

The climbing vines of which I have spoken were scant and puny as they receded from the vicinity of the opening. There can be little if any vigorous vegetable growth without sunlight, and little as was the quantity that entered the chamber, it exerted a marked influence in its immediate neighborhood.

But the search, prosecuted as it was under many difficulties, satisfied Hermon that if he ever left the place it must be through the same passage which admitted him.

"And how can this be done?"

This was the question which finally brought him to a halt near the fire, with his eyes turned upward toward the opening, as if he expected to read the answer there.

"I can never reach that without help, that's certain," was his truthful conclusion; "and where is the help to come from?"

He knew that as a matter of course he would be missed in the morning, and Jurak and Eustace could be counted on to do everything possible to find him. They could not track him unaided, and he felt that it was doubtful whether any dog could be found that would do it. Tweak being left out of the calculation, he had little belief, from what he had seen of the native canines, that they would prove any better.

The youth's manner of entrance was such that he could know little of the outward appearance of the hole in the ground, but it seemed that a diligent hunt on the part of his friends and others ought to find the spot. Had he known the truth, he would have been less hopeful of that means of escape.

CHAPTER XXIX.

WALLED IN ON EVERY HAND.

IT IS hard for a strong, active boy, placed in a situation like that of Hermon Hadley, to fold his arms and sit down with the conviction that he can do nothing to help himself. While our young friend did not exactly do this, yet he was forced to that state of mind.

One dread was with him while making his circuit of the chamber—he expected to come in contact with some of the deadly serpents found in Java. It seemed to him that it was just the place for them; but when he returned to the fire he was quite sure there were none near him. He had seen and heard nothing of them, and the warmth from the fire would have drawn them from their hiding-places. It was a great relief to believe he was not to share the place with such unwelcome neighbors.

Hermon found his watch had not been injured by his fall into the chamber, and he was therefore able to note the passage of time. When he looked at its face he saw that it lacked a couple of hours of daylight.

"They won't miss me till morning," was his conclusion, "and then it will be some time before they'll think anything is wrong. Pretty soon they'll begin to hunt, and some of them ought to pass near that opening up there. About that time I'll begin to make a noise."

The youth felt that sleep was out of the question. His body still ached, and he could not help asking himself the question, "Will they find me?"

The dread that they might fail to do so made it impossible to keep still for any length of time. He sat down by the fire and looked into the embers while he thought and thought. Then, starting up, he walked back and forth through the dimly-lit chamber. He occupied himself in gathering all the fuel he could, so that he might keep his camp-fire, as he regarded it, going as long as possible. Now that there was a good bed of coals it was not necessary that the wood should be so dry, and he collected everything obtainable.

With the help of his knife he loosened the tiny but innumerable tendrils which held the vines in place and pulled down yards of them at a time. Often, when the mass of branches and leaves came tumbling about his head and shoulders, he started with the fear that among them was a serpent striving to wind itself about his neck or to strike him; but in every instance he was relieved to find no cause for such fear.

Hermon continued work until it seemed to him there was little left to gather, and he had enough to keep the fire going several days. Heaped near the blaze on the flinty floor, it made the best kind of a bed, on which he threw himself. The feeling was so grateful, and his weariness had been so increased by his labor, that unexpectedly to himself he sank into a sleep which lasted several hours.

When he opened his eyes and recalled his situation he saw the fire was down to a few embers so covered with ashes that everything was dark around him. His first act was to stir these embers and to throw on more fuel. This quickly crackled and blazed, until once more the whole interior was illuminated. Then, when he looked at his watch, he saw that the sun had been shining on the earth above for more than two hours.

"They're at it now," said Hermon, thinking of the alarm that Jurak and Eustace must be feeling at that moment over his disappearance. "I shouldn't wonder if they're not far from the opening, and if they can only find a dog that will track me it won't take long to discover what hole I went through."

He now began hallooing with might and main, and kept it up until hoarse. Then, with the aid of his tongue and fingers, he sent out a blast which reverberated from end to end of the chamber until his ears tingled. This was done again and again, after which he drew his revolver, with the intention of using up most of the cartridges of that. But he had been less fortunate with his pistol than with his watch. The hammer was injured in such a manner that it could not be fired until after repairing.

"I don't know that it makes any difference," he concluded, returning the weapon to its place, "for the report won't travel as far as the whistle, which Eustace will know, provided he hears it."

Having done all that was possible in the way of signaling, he moved back and listened for some evidence that he had been heard. He was not very hopeful of success, and minute after minute passed with the profound stillness unbroken. If any people traversed the flinty roof overhead, the sound of their feet was so muffled that not the least indication reached the straining ear of the listener.

But while gazing upward Hermon was able to distinguish very faintly the irregular outlines of the opening through which he had descended. This was proof that the sun was shining above; and though the grass and bushes shrouded the entrance, yet enough of the sunlight penetrated them to reveal the place.

"That's it, sure enough," was his conclusion. He understood, too, why the outlines were so dimly revealed. "That looks bad for me, because it will be hard for them to find the spot."

Then he resorted to his signaling again, keeping it up until he became so tired he was forced to stop.

In this manner the hours dragged slowly by until

his watch showed the afternoon was half gone. The fire was still burning, though with less vigor, for he was gradually reaching the belief that even if delivered from his peril he would have to stay a long time, and he could not bear the thought of being shrouded in the impenetrable darkness which would otherwise fill the place. The fire, therefore, was permitted to burn so low that the chamber was only partly illuminated, and the darkness beyond seemed the more profound and impressive from the contrast.

CHAPTER XXX.

AN UNEXPECTED FRIEND.

HERMON HADLEY carefully rewound his watch and looked intently upward, but was unable to distinguish the outlines of the opening which had admitted him to the chamber. It was night, and darkness reigned above as it did beneath the earth.

Beside the question which was ever present with the youth, there was another that could not be put aside. He had been twenty-four hours without drink or food, and so long as he remained in the chamber it was impossible to procure a particle of either. The only thing in the way of vegetation to be found was the growth of vines. These bore nothing in the shape of fruit, and had they done so he would have been afraid of eating it, through dread of poisoning. Anxious and frightened as he was he could not keep off the gnawing hunger and thirst, which steadily grew in their intensity.

As we all know, thirst is much more distressing than hunger; and this is especially true of warm countries, where the want of water is quickly felt. One end of the chamber seemed to be more damp and moist than elsewhere, but it was not enough so to exude a drop of water.

Loth to admit his helplessness, Hermon rarely remained on the pile of vines, except when asleep, more than a few minutes at a time. Then he would rise and walk around his prison, peering here and there with the idea of finding some peg on which to hang a slight hope. When the hour grew late, and he thought he could sleep, he stirred the flames to greater vigor than ever and started on a last round, as it may be called.

He was approaching the furthest corner, away over on the west, when he discovered something that he had not noticed before. He could not see it distinctly, but suspecting its nature he hastened to the fire, picked up the largest brand, and circling it over his head until it broke into a vigorous blaze walked back.

There in the corner lay what had once been a living person like himself. The clothing was perfect, but he who had worn it had long since crumbled away until only a skeleton remained.

There could be no mistaking the identity of the remains; they were those of the youth who disappeared ten years before, and of whom no trace was ever found. He had wandered off from his friends as had Hermon, and the same strange fatality had caused him to fall through the opening which precipitated the young American into his dismal prison.

Hermon paused only long enough to take a hasty survey of the sad sight, when he walked to the fire and threw his torch upon the flames.

"It is the strangest thing I ever knew that I should follow in the footsteps of that poor fellow. His friends, Jurak told us, hunted long after all hope seemed gone, and there was nothing that could be done which they did not do; and yet he was left to die here, as I also may be left to die."

The unfortunate lad must have fallen into the chamber in the same manner as had Hermon, for had he been killed, his remains would not have been found where they were. He may have been badly wounded.

It was a great shock to the young American, and when he reached the pile he sat down in a more despairing frame of mind than he had felt since his misfortune.

"Can it be that I, too, am to die here?" he whispered, looking furtively around. "Must Jurak and Eustace go back home and say I was lost and no one could find me? Is the opening through which I fell so hidden from sight that it cannot be found by the most careful searching? It would seem so from what Jurak told us; but these ruins are so well known that some of the natives ought to have learned all about the under-ground chambers."

The hour was late, but there was no sensation of drowsiness to close the eyes of Hermon Hadley.

All such feeling was driven away by the sad discovery just made. Brave as he was, he was unnerved for the time by it.

Sitting in this disconsolate mood, he paid no heed to the fact that the fire was burning low. Looking gloomily into the faint embers, all at once he became aware of a slight noise overhead near the opening. Starting to his feet, he glanced aloft and called out:

"Halloo! who's there? Is that you, Jurak, or Eustace?"

He listened, but no reply came to him.

"It must be some animal," was his thought; and then he shouted in a louder voice than before:

"Halloo, up there! Who are you?"

He heard, by way of answer, the short bark of a dog.

"My gracious!" he exclaimed, "I believe that is Tweak!"

CHAPTER XXXI.

RESCUED.

HERMON HADLEY held his breath. Again he heard that peculiar cat-like bark with which he had become familiar on his jaunt through Java. It was the dog Tweak. Aye, indeed; there could be no mistake about it.

"Halloo, Tweak!" shouted the youth. "If that's you, and you'll do all you can to help get me out of this place, I'll beg your pardon for all the mean things I ever said about you!"

The canine seemed to be pawing at the opening as though the boy were a squirrel which he was digging out. He gave utterance to the short, quick barks which his species often emit at such times, and Hermon continually called to him, so that the brute might feel sure no error was made in the matter.

The lad threw more fuel on the flames until they blazed high and the interior of the huge chamber was fully illuminated. Then he peered upward, in the hope of catching sight of the animal. Once he fancied he saw the sparkle of his eyes as he thrust his nose downward and snuffed and clawed, but more than likely he was mistaken.

"Say, old dog. I'll help put you on a pension for life if you'll stick to it till Eustace and Jurak come!"

Again he listened, but all was still. He shouted to the canine, whom he could have hugged for gratitude; but the bark was silent.

"I suppose he's tired and is resting awhile. Now if, instead of hunting for me, my friends will search for Tweak, they will accomplish something."

It was the second night of Hermon's imprisonment, and the hour was so late that he did not believe any help could come to him before the morrow. It was not likely his friends would continue their search during the darkness, and it would take some time after the rising of the sun to trace him.

"I may as well try to sleep," was his conclusion, as he threw himself on the pile of vines, which had been very much reduced in size. He was now in such a hopeful frame of mind that he felt it possible to sleep.

"*Halloo! Hermon, old fellow!*"

The youth was sinking into drowsiness when he was brought to his feet by the voice of his cousin Eustace, who shouted down the opening. The elder, as the reader will recall, had been attracted by the actions of Tweak when he approached the camp-fire, and followed him through the moonlight until the canine stopped beside the bush which concealed the entrance to the subterranean chamber. The first thought of Eustace was that the dead body of his friend lay beneath, but he carefully investigated. Guarded as he was in his movements, he missed by a hair's-breadth shooting down the opening, just as his cousin had done.

Catching the bushes and recoiling just in time, he called the name of his cousin; and the latter was not more startled by the hail than was the elder when the cheery response came back:

"It is I, Eustace! Sound in body, but thirsty, hungry, and anxious to get out!"

"Well, have patience and we'll give you a lift."

The delight of the lads was so great that they could not help chaffing each other for a few minutes.

"Can't you drop in and see a fellow, Eustace?"

"I came near doing so a minute ago but changed my mind. When I call on a person, I prefer to go into his parlor rather than his cellar."

"So do I; but when a fellow leaves his cellar-door open there is no telling who will fall into it. My entrance, however, was made in the best style. Where's Jurak?"

"Waiting down by the old temple, two or three hundred yards away."

"How did you learn I was here?"

"Tweak came to camp and acted so queerly that I knew something was up, and I followed him."

"Is he there with you?"

"Yes."

"I wish you would apologize to him for all the

slighting remarks I have made. I take them all back, and henceforth shall speak of him in the most respectful terms."

Hermon felt a kink in his neck from looking upward so much, and he now lay down on his back on the pile and called to his cousin, asking him how he proposed to help him.

"We shall have to get a rope and let it down to you. How deep is it?"

"About fifty feet, as near as I can guess."

"Well, I will now go back to Jurak and explain it all to him. He will have to go to the village for the rope, and we may have to wait till morning."

"All right; do what you think best."

Eustace called good-night, and with a glad heart set out to acquaint Jurak with the joyful news. Tweak trotted in advance, no doubt conscious of the good deed he had done and the vast advance he had made in the estimation of his acquaintances.

Before the fire was reached he met the native, to whom he told the news. Jurak was not excitable, but he sprang from the ground, struck his thigh with the flat of his hand, and uttered a fervent ejaculation of thankfulness. He started to hasten to the spot, but changed his mind and hurried to the village. The hour was so late and he found so much difficulty in procuring what he wanted that he finally decided to wait till morning. He therefore rejoined Eustace, and the two lay down by the fire, talking long and earnestly together. The native gave it as his opinion that the missionary's son had lost his life from a similar misfortune, but as Hermon had not yet made known his discovery, it remained a conjecture for the time.

In the meantime Hermon had fallen asleep, though it was almost daylight when he did so, and he did not open his eyes until aroused by the shouts from above. He sprang to his feet, called back that he was ready and waiting, and then for the last time renewed the smouldering fire until once more it filled the chamber with light.

Looking aloft toward the dimly-outlined opening, he saw the end of a rope dangling and descending like an attenuated serpent. He waited patiently until it came within reach, when he seized it, and when enough more was at command coiled it about his body, then around his arms, and shouted:

"Now show how much muscle you've got up there! My gracious!"

The words were hardly uttered when he was lifted a dozen feet with great swiftness, and continued to ascend with such speed that he shouted to them to be careful or they would fracture his head against the rocks above. Thereupon Eustace, who was directing operations, allowed him to dangle awhile like a huge pendulum, until Hermon inquired with a yell what the matter could be.

Up again he went, and before he was aware of it

the mountain seemed to cast him forth, and he fell prostrate on the earth at the feet of Eustace, Jurak and a dozen natives, who had been pulling at the rope. The rescued lad could hardly see in the glare of light (because he had been so long in the darkness), and despite his sturdy frame was so overcome that for several minutes he was too weak to stand.

But every preparation had been made for him. Good, cool water and delicious food were there in abundance, and he ate as an American boy may be expected to do when he has been a long time on short allowance.

Hermon was in such a plight, especially as regarded his clothing, that he and Eustace went to the house of Marriavo in the village to recruit and have his garments repaired. While there, Hermon told the story of the body which still lay in the chamber where he had been imprisoned more than twenty-four hours.

His narration, it need not be saio, produced a profound impression. Marriavo asked many questions, and told the lad he ought to have searched the clothing for evidence, or rather mementos, that could be returned to his friends. The good-hearted but somewhat cranky host was compelled to give up his theory that marauding sailors had carried away the boys, since the evidence was too direct to be gainsaid; but he insisted that the evil pirates had been prowling through the village, and that they would have abducted Hermon had he not fallen into the subterranean chamber just in time to save himself from such a fate.

There was no one in the settlement that could repair the rifle and pistol of Hermon, but the lad was in such high spirits that he insisted they should start homeward without delay, with the intention of doing considerable hunting on the road. He consented, however, that horses should be procured, and the return trip was begun under more advantageous terms than was the jaunt from home. As good fortune would have it, they had not gone far when Jurak was able to borrow a weapon from an acquaintance.

The first point at which they made a halt worth noting was near the coffee plantation where Hermon had shot the leopard. True to the promise made him, the skin was awaiting his return. It had been stretched in the sun, and though not perfectly dry was in condition to take away. He distributed a number of presents among his friends, thanked them kindly, and the little party continued northward toward home.

When the most mountainous region was reached the desire of the boys for another hunt became so strong that they left their animals with a friend, and the three entered a deep forest for the purpose of engaging in their favorite sport.

"This looks like a good gun," said Hermon, surveying his weapon doubtingly, "but I don't believe it is as good as my own was."

"That's an old excuse for poor marksmanship," said Eustace. "A sportsman is like every one else— he will lay the cause of failure to anything except his own blundering stupidity."

"I don't know that I have had any more failures than a certain young gentleman I could name," replied Hermon, too happy to feel anything like resentment at the good-natured chaffing of his cousin, whom he could not have loved more had he been his own brother. "The fact is, I think I was a little ahead on the way out."

"Oh! some persons are born lucky," airily responded Eustace; "but it can't last forever."

"No; for if it did, you might gain a part of the skill of an ordinary hunter. Wait till we have a fair chance."

"There it is!" exclaimed his companion, pointing to a bird sitting on a limb not more than fifty feet away.

"What sort of an animal is that?" asked Hermon, staring at the object, while he cautiously made his weapon ready.

Eustace asked Jurak and then replied:

"It is called a lorikeet."

"That's what I thought," remarked Hermon, who had never heard the name before. "Now see him vanish as though he never existed."

Bringing his gun to a level, he sighted carefully, while the others watched the result with breathless interest.

Bang! and the bird never stirred.

Eustace and Jurak broke into hearty laughter, as much at the look of dismay on the countenance of the young American as at his woful failure.

Then, as soon as Eustace could control his mirth, he brought his own weapon to his shoulder and let fly. The bird whizzed through the leaves and dropped to the ground as if struck by a grape-shot.

"That was a mean piece of business," said Hermon. "After I put a bullet right through his body, killing him so quickly that he didn't know it, you must wake him up with the noise of your gun. I shouldn't be surprised if you would claim that you killed him!"

And Hermon stared at his laughing companion with an assumption of anger that deceived no one.

"Oh, no," said the latter, his face becoming serious; "I merely roused the bird to the fact that he was dead—that's all."

"Very well; don't go to getting up any other yarns when you come to tell your father about our trip, after we reach home."

Late that afternoon, when the hunters were looking about for an inviting place to camp, a kubin or flying cat suddenly skurried from almost under their feet. Hermon hastily aimed and fired. The animal was hit, but would have got away had not the youth dashed forward and finished it with the butt of his weapon.

Eustace looked very serious.

"Hermon, why are you so extravagant?"

"What do you mean?"

"You shouldn't waste your ammunition in that reckless style. Why don't you run up and whack over the head whatever game you see, and save your cartridges? If you can't catch it, Jurak and I will hold it for you!"

"That wouldn't do, for I would be likely to mistake you for some of the monkeys and slay you instead of the game!"

"You could avoid that by aiming at *us;* then we would be safe!"

"If this gun continues to play me such tricks I'll cast it aside and throw stones at whatever I see!"

"There it is again!" exclaimed Eustace with a sniff. "You are laying it to the gun instead of owning up like a man!"

"If you think it is not the fault of the weapon, suppose you trade weapons and try mine?"

"I am afraid I shall have to do it if matters go on this way much longer," said Eustace.

"There's one thing certain: if this gun don't reform I shall withdraw from the syndicate and leave the whole business to you!"

A half hour later they struck the very spot for

which they were looking. It was in the depth of the forest, where a small stream of clear running water flowed, and where the vegetation was so dense that no sudden swirl of the elements could disturb them. Jurak had been thoughtful enough to bring a substantial lunch with them, on which they not only made an excellent supper but had enough left for the morning meal

It was agreed that Hermon should keep watch the early part of the night, while Jurak would take his place about "low twelve" and keep guard till daybreak. The three stretched themselves out before the cheerful blaze, and talked a long time over the wonderful experiences that had befallen Hermon in the subterranean chamber, where, for a time, he believed he was doomed to perish like the unfortunate lad who was overtaken by a similar misfortune many years before.

The youth related the particulars of his descent into the frightful place, and spoke of the lonely, dismal hours which were hardly cheered by the fire he was enabled to kindle.

"Tweak," he said, drawing the mongrel toward him and patting his frowzy head, "I have apologized already for the bad things I said about you. I don't suppose there is any use of pretending that you are the bravest dog that ever lived, but you have earned my gratitude and proven that you are like us—no matter how contemptible we may seem, there is some good in us which is only awaiting the opportunity to show itself."

And he patted the dog's head affectionately, while the canine, as if conscious of the exploit he had performed, wagged his tail and walked about with the air of one who understands the obligations under which he has placed others.

Finally, as all felt weary from their long tramp through the woods, Eustace and the native wrapped themselves in their blankets and were soon asleep, leaving the camp in charge of the sturdy young American, who felt himself equal to the duty, which he had performed more than once before while on their memorable jaunt through Java.

"There are plenty of wild animals in this part of the world," he reflected, when he realized that he was alone, "and if I ain't mistaken there are some of them prowling around the camp this very minute."

The cause of this conclusion was not that he saw or heard anything, but the actions of Tweak. The pup may have wished to assume a brave demeanor, so as to maintain the reputation he had gained from his recent achievement; but really the poor creature was not equal to the demand.

He gave utterance to several threatening growls, bristled up, and stared out in the darkness in a way which proved that his keen intelligence had scented some danger there. Then, at an encouraging word from Hermon, he darted out in the gloom as if fully resolved to chew up the threatening brute, even if it was a tiger; but he had taken only a half dozen steps when he came back as if fired from a catapult.

"Don't be scared," said the sentinel. "We are here with you, and no matter what it is, you can count on us to help you out of any scrape that you may get into."

Once more Tweak plucked up courage and stepped guardedly into the encircling gloom, but got no further than before when he skurried back and sought refuge behind Hermon, who laughed in spite of himself.

This time there was the best reason for the frantic retreat of the animal, for he was pursued by something which made a desperate attempt to seize him, and came very near succeeding.

Remembering the unreliability of the gun he had been using through the day, the lad had stealthily extracted his cousin's weapon from his embrace and substituted his own without awaking the sleeper. It was the fact that the young American held a rifle on which he could rely that gave him so much confidence.

The enemy which pursued Tweak so fiercely was a lutung, a species of ape that sometimes shows audacious bravery, as he did on the present occasion, though it must be admitted that he also displayed a lack of judgment; for instead of halting when the canine took refuge behind his master, whining with terror, the lutung continued to advance, not only with the manifest resolve of killing the little brute, but of attacking Hermon as well.

"If you had enough sense to stay away," said the lad to himself, "I wouldn't disturb you; but you began the row and I will end it."

Thereupon he sent a bullet through the skull of the ape, which leaped convulsively in the air and fell dead so close to the sleeping couple that his forearms dropped across the feet of Eustace.

"My gracious! what's the matter?" demanded the latter, leaping up at the same moment that Jurak sprang to his feet and grasped his gun.

"There's another one of them!" shouted Hermon to his cousin; "fire quick or he will get away!"

Jurak was on the point of discharging his weapon when Eustace demanded that it be left to him to dispose of.

Hermon smiled to himself, for his cousin had not noticed, in his excitement, that it was not his own weapon which he held.

The second lutung, probably a mate of the other, stole forward in a way calculated to awe the bravest hunter, for there was something so suggestive in the slow, sure and noiseless approach that the sportsman could not help feeling like the man who stands in front of a gorilla. If he misses, he is sure to be torn to pieces before he can fire a second shot.

Awaking from a sound sleep, the lad was naturally flurried; but the distance was so slight that he could not have missed with a weapon which was good for anything at all.

Eustace took but a moment to aim, when he pulled trigger. The lutung started slightly, as if from the flash of the rifle in his face, but he continued to approach as before.

"I've missed! I've missed!" shouted the youth, leaping back in affright. "Quick, Jurak!"

The latter let fly, and the lutung was extinguished with the same suddenness as his predecessor.

It was Hermou's turn now, and you may be sure he did not spare his cousin.

"Why didn't you wait, Eustace, till Jurak and I caught the creature and held it still? Then you could have shoved the muzzle in his ear, and I don't think you would have missed!"

"I don't understand how it was," replied the other, looking around with a confused air, as though the whole thing was beyond his comprehension. "I'm sure I aimed well enough."

"Some folks will lay their failure to their guns, instead of owning up that they don't know anything about shooting!"

Just then the native pointed at the rifle in the hands of Eustace, who looked down and discovered the trick played on him.

"Ah," he laughed, "you have it on me this time! I own up!"

After this demonstration of the worthlessness of the borrowed weapon, Hermon expressed his desire to bring the hunt to an end. Since it was out of his power to bear a creditable part in the sport he had no heart left. Beside, they had been away from home so long that all felt a natural desire to reach there without unnecessary delay. It was agreed, therefore, that the return should be made as quickly as possible.

Accordingly the next morning they made their way out of the forest to the highway and pushed on till they reached the house of Jurak, where Myeta warmly welcomed them. She had saved the beautiful tiger-skin for Hermon, and the lad was delighted. Taking his cousin aside he inquired how much funds he had left, and when told borrowed every cent of it, and adding about all his own he presented it to Jurak and his wife, who were made as happy as a couple of children.

A few days later, and without any incident worth mentioning, the boys reached the home of Eustace, and their ever-memorable jaunt through Java was ended. Some weeks later they sailed for England in company with the mother of Hermon, and the two entered one of the most famous institutions of learning in the kingdom. And there we say good-by and part company with them.

THE END.

www.ingramcontent.com/pod-product-compliance
Lightning Source LLC
Chambersburg PA
CBHW032146230426
43672CB00011B/2466